Smart Glass Technology

Smart Glass Technology

By

Mujtaba Ellahi Mahar, Shaista Taimur
and Hamdullah Khan Tareen

Cambridge Scholars Publishing

Smart Glass Technology

By Mujtaba Ellahi Mahar, Shaista Taimur and Hamdullah Khan Tareen

This book first published 2025

Cambridge Scholars Publishing

Lady Stephenson Library, Newcastle upon Tyne, NE6 2PA, UK

British Library Cataloguing in Publication Data
A catalogue record for this book is available from the British Library

ISBN: 978-1-0364-5009-0
ISBN (Ebook): 978-1-0364-5010-6

I would like to dedicate my book to my beloved parents, especially to my late father and mother.

TABLE OF CONTENTS

LIST OF TABLES

LIST OF FIGURES

LIST OF ACRONYMS/ABBREVIATIONS

LC	Liquid Crystal
LCD	Liquid Crystal Display
PDLC	Polymer Dispersed Liquid Crystal
TIPS	Thermally Induced Phase Separation
SIPS	Solvent Induced Phase Separation
PIPS	Polymerization Induced Phase Separation
NCAP	Nematic Curvilinear Aligned Phase
ITO	Indium Tin Oxide
$\varepsilon\|$	Dielectric anisotropy parallel to the director
$\varepsilon\perp$	Dielectric anisotropy vertical to the director
$\Delta\varepsilon$	Dielectric anisotropy
no	Ordinary refractive index
ne	Extra ordinary refractive index
np	Refractive index of the polymer
neff	Effective refractive index
Δn	Birefringence
n	Director
Toff	Off state transmittance
Ton	On state transmittance
V_{th}	Threshold Voltage
V*sat*	Saturation Voltage
CR	Contrast Ratio
τ_R	Rise time
τ_D	Decay time
SEM	Scanning Electron Microscopy
POM	Polarize Optical Microscopy
LCT	LCD parameters tester
AgNPs	Silver Nanoparticles
RI	Refractive Index

CHAPTER 1

INTRODUCTION

Abstract

Smart glass technology has become increasingly prevalent in modern applications. One prominent type is a polymer-dispersed liquid crystal (PDLC), which consists of epoxy monomers with liquid crystals dispersed throughout. By incorporating silver nanoparticles in varying concentrations, PDLC can achieve different sizes of nanometer distributions, enhancing its properties. The characterization of these silver nanoparticles is typically conducted using UV-visible spectroscopy, with sodium borohydride used as a reducing agent to control their size. The electro-optical properties of PDLC are analyzed with a liquid crystal detector and an Abbe refractometer for refractive index measurements. Compared to traditional LCD materials, PDLC thin film technology offers numerous advantages, such as simplified manufacturing processes, elimination of polarizers, and mitigation of viewing angle limitations. Additionally, it does not require precise control over the spaces between substrates, making it suitable for large-area displays. The production of thin PDLC films is achieved through a method known as polymerization-induced phase separation (PIPS), using mixtures of curing monomers, liquid crystals, hardeners, and silver nanoparticles. Research indicates that the inclusion of silver nanoparticles at various sizes significantly influences the polymer network's structure, particularly increasing the crosslinking density when combined with curing monomers that contain -NH2 groups in epoxy monomer/liquid crystal/hardener mixtures. This allows the film to transition from a transparent to an opaque state, enabling a broader range of applications in PDLC technology and other optical devices.

The "smart" windows are used for the construction of a smart building as privacy glass. It is used for decreasing the energy usage in buildings. Energy in the form of radiation and heat is used as stimuli for smart materials the smart material can control the transmission of light in smart materials consisting of liquid crystal used as a matrix. Smart display devices are used as privacy windows and energy-saving windows (Castellón & Levy, 2018). The display devices might have some electric and optical signal issues that contain an uneven distribution of electric fields non-uniform optical behavior has occurred at a large area of displays. To overcome these issues more efforts are required.

1.1 Smart Glass technology

Smart glass technology is an advanced glass technology that is used for privacy glass, switchable glass, magic glass, and smart windows. In three decades, three generations of smart glass were discovered and used by retailers. First-generation (1G) used emulsion technology, in which water-soluble polyvinyl alcohol is dispersed in liquid crystal droplets. The second-generation (2G) is composed of epoxy dispersed in droplets of liquid crystal known as polymer dispersed liquid crystal PDLC technology and the third-generation (3G) synthesizes by a nonlinear polymer to disperse in a liquid crystal called non-linear PDLC technology.

The advancements from 1G to 2G demonstrated improvements in features, but 3G technology significantly enhanced usability for both indoor and outdoor applications. This generation stands out as it provides better electro-optical properties and higher-quality performance compared to its predecessors. As a result, 3G has become more widely adopted, showcasing its effectiveness in various environments (Pandian, Karthikeyan, Rajasimman, & Dinesh, 2015).

1.2 Historical view of Polymer Liquid Crystal Composites

Since the beginning of life, polymers and liquid crystal (LC) materials have been integral to biological processes. The structures of Deoxyribonucleic acids (DNA) exhibit polymer-like characteristics. Lyto-tropic liquid crystals made up of macromolecules display various LC phases. Cell membranes consist of atoms, featuring hydrophilic liquid heads and hydrophobic organic compound tails. Naturally occurring plant polymers include rubber-like polyose and resin. Despite their widespread presence, liquid crystals were not discovered until the late nineteenth century due to the heating of plant materials. In the mid-1900s, the development of plastic emerged through the process of polymerization, specifically using phenol (C6H5-OH) and formaldehyde (CHOH), leading to the synthesis of more complex macromolecular compounds (Carothers, 1929). Numerous explorations have been conducted on solid-liquid interactions and compounds, although the application of these materials has been performed in various ways. Fluid gems initially evolved as dynamic displays, temperature sensors, and novelty toys. Polymers have been integrated for a long time, serving as traditional elastic alternatives and plastics. The primary usage of polymers in fluid jewellery lies in their arrangement on substrates, as the chemical compound chains influence the direction and alignment of the fluid gem particles.

In the 1980s, significant advancements were made in the study of intermediate solid-liquid substance composite compounds. A notable figure, James Fergason, co-designed the aromatic nematic fluid jewellery display in 1984. He elaborated on two key types of compound-dispersed liquid crystal films (PDLC). These films were characterized by being sandwiched between two conductive substrates, with liquid crystal materials mixed with polyvinyl alcohol (PVA). This innovation paved the way for the development and application of fluid gems in various technologies (Fergason, 1985). Moreover, in 1985, William Doane and his colleagues at the Liquid Crystal Institute at Kent State University further advanced the development of PDLC films. They innovated a method that involved blending epoxy with liquid crystals, allowing the two materials to coexist independently. This approach facilitated the creation of more versatile and effective PDLC films, contributing to the ongoing evolution of liquid crystal technology and its applications. (Drzaic & Drzaic, 2006).

Thereafter, when an electric field is applied to the dual-stage substances, the material transitions from a hazy state in the absence of an electric field to a clear state when the field is activated. From that point on, there has been a wide range of chemical compounds that have evolved, varying from simple mixtures to more complex formulations. Additionally, the use of resins in liquid crystal applications has become notably significant. The polymer matrix remains intact, encasing the liquid crystal droplets, which allows for switchable light scattering. Furthermore, the polymer matrix also acts as a stabilizer, enabling the liquid crystal to maintain its desired phase or to facilitate a specific liquid crystal state (J. Doane, Vaz, Wu, & Žumer, 1986). Indeed, even fluid gem liquids could be formed into gel or solid with receptive to improvements (i.e., light, Temperature) when polymeric substances are available.

1.3 Introduction to PDLC

PDLC, or Polymer-Dispersed Liquid Crystal, is a material that combines liquid crystals with lower molecular weight and a polymer matrix. This is achieved by creating micro-sized transparent solid droplets dispersed within the polymer. PDLC finds applications in various fields including smart displays, privacy glass, high-contrast reflective displays, and projection displays. These materials are typically used as thin films for a range of products as switchable windows, privacy glass, large advertising screens, and television displays. The versatility of smart displays highlights the wide-ranging commercial applications of this technology in today's market. There are two main types of privacy glasses: the first is a polymer-dispersed liquid crystal (PDLC) the second is a polymer-stabilized liquid crystal (PSLC), each offering unique properties suited to different uses (Mormile, Musto, Petti, Ragosta, & Villano, 2000).

Polymer Dispersed Liquid Crystal (PDLC) techniques have already found their applications in various advanced technologies, including optical communication, 3D screens, switchable zoom lenses, and wavelength division multiplexing waveguides. These applications leverage the unique properties of PDLC materials to enhance functionality and performance in visual displays and communication systems, allowing for innovative solutions in both consumer electronics and professional settings (Liu, Zheng, Shen, Wang, & Zhuang, 2019). PDLC devices have gained significant attention over the years due to their numerous benefits, making them increasingly popular. The PDLC technology continues to evolve, with advancements such as the incorporation of surfactants, cross-linking agents, and doping additives, all aimed at enhancing its reliability and effectiveness for various applications (M. Zhang et al., 2013).

In PDLC (polymer dispersed liquid crystal) technology, the liquid crystal is crucial for display preparation, particularly concerning its electro-optical properties. The droplets of liquid crystal contribute unique electro-optical and magneto-optic characteristics to PDLCs, enhancing their functionalities. However, since liquid crystal is a costly material, minimizing its content can significantly reduce the overall cost of the PDLC displays, ultimately making them more affordable for consumers. This balance between performance and cost is key to making PDLC technology more accessible while maintaining its desirable features (Kashima et al., 2010). PDLC exhibits a unique property where it becomes transparent in the off-state and appears milky when activated in the on-state. Currently, Liquid Crystal Displays (LCDs) are undergoing rapid development, making them a prominent technology within the industry and a significant factor in economic growth. The liquid crystal sector is crucial in the polymer field, continuously evolving to meet rising consumer demands. However, as display materials improve, there is a growing expectation for advancements that address existing limitations, such as slow and lagging transmittance and the need for high drive voltages. These enhancements are essential for meeting modern requirements in display technology (Chen, Chen, Sun, & Wu, 2006).

1.4 Polymer Dispersed Liquid Crystal Formation

A homogeneous tropic solution of polymers and pre-polymers is prepared with a lower molecular weight of liquid crystalline substance via way of means of the separation method. The boom takes place with the dissolved liquid crystal in solution liquid crystal section separates and droplet formation is achieved. The instruction mechanisms are solvent-induce phase separation (SIPS), thermal induced induce phase separation (TIPS) and polymerization-induced phase separation (PIPS) methods. The main principle is illustrated in every method and applies to the TIPS, PIPS, and SIPS methods.

In the 1980s, PDLC (Polymer Dispersed Liquid Crystal) was developed using the phase separation method, which involved using UV-curable epoxy pre-polymers and thermoplastic polymers for its preparation. Various films of PDLC materials have been produced, showcasing a range of shapes and thicknesses. The predominant technique for preparing these films involves ultraviolet polymerization, specifically the phase separation method (PIPS). This approach facilitates the creation of a polymer network within the PDLC film, influencing its optical properties and overall performance (Hakemi, Khenkin, Tamari, & Gal-Fuss, 2020). PDLC (Polymer Dispersed Liquid Crystal) technology has significant applications in large-area displays and high-definition displays, making it a promising option for projection TVs. This technology offers several advantages, including a wide viewing angle and the ability to easily enlarge the display area. These features contribute to its versatility and attractiveness in various display applications, enhancing both functionality and user experience.

1.4.1Thermal-Induced Phase Separation (TIPS) Method

This separation method uses polymers that do not crosslink and melt at reasonably low temperatures The TIPS method is which that melts below their decomposition temperatures. This method is utilized for thermoplastics and is called as TIPS method.

1.4.2 Solvent induced Phase Separation (SIPS) Method

Initially, in the SIPS method, polymer and crystal liquid form a similar solution with a common solvent. The solvent is then evaporated, thus bringing the solution into the immiscible region. Although the liquid crystal phase separates, droplets grow and appear through polymer solidification (K. Li et al., 2020). The rate of evaporation controls the droplet size. This method of separation is called as SIPS method.

1.4.3 Polymerization induced Phase Separation (PIPS) Method

PIPS method is convenient for pre-polymer materials that are miscible with liquid crystalline phases. Depending on the polymer precursors used, the PIPS method could be either photo-initiated or thermal-initiated. For photo-initiation, reliable PDLCs mostly undergo acrylate or thiol-ene chemistry. These are polymerized by an addition reaction for thermal initiation. The PIPS method can produce PDLC films with droplets less than 10-8 nm.

Phase-separated composite film and PDLC synthesized by polymerization separation (PIPS) technique could be used for the study of the optical function of composite materials. The (PIPS) method is widely used because it's solvent-free, inexpensive, eco-friendly, less contaminant, and controllable. That's why it is mostly used to produce PDLC films (Saeed et al., 2021). Polymerization Phase partition (PIPS) by and large uses polymers framed by a stage development expansion polymerization response. A low-sub-atomic weight fluid liquid crystal is disintegrated in a pre-polymer arrangement. Polymer formation happens either by temperature or photographs artificially, variably the synthetic capability of arrangement and lessening the dissolvability of fluid precious stone. The arrangement goes through the miscibility hole the fluid gem stage isolates into beads.

Stage partition can occur through shaft disintegration or bead nucleation and growth, particularly in the context of liquid crystals of low-molecular-weight distributed within a polymer. The study of polymeric materials has captivated researchers for many years due to their diverse applications. Epoxy, in particular, stands out as a polymeric material noted for its excellent wetting properties, mechanical strength, chemical resistance, and fire retardance, making it suitable for a wide range of industries and applications (Paluvai, Mohanty, & Nayak, 2014). Epoxy resins are indeed crucial in the fields of transportation, construction, and communication. Recent advancements in epoxy resin technology have enhanced their properties, making them even more valuable and reliable. These improvements include increased durability, enhanced adhesion, and better resistance to environmental factors. As a result, epoxy resins are increasingly used in high-performance applications, such as in the manufacturing of composite materials for vehicles, protective coatings for infrastructure, and adhesives in electronic components. This evolution not only expands their applicability but also improves the overall effectiveness and longevity of the products in which they are used (Mohan, 2013).

The diverse usage of epoxy resin is prevalent in applications such as Aeroplanes and wind turbines, where it is utilized as a composite in various industries. Epoxy is favoured for its engineering purposes, adhesive properties, high chemical resistance, and use in 3D printer ink, and it also offers recyclable qualities. When compared to traditional display modes, the advantages of PDLC (Polymer Dispersed Liquid Crystal) technology are significant: it does not require polarizers or specialized substrates, and the surface orientation treatment is unnecessary. Additionally, the spacing between the two substrates can be less controlled, resulting in a product that is thinner, lighter, and exhibits higher brightness (Capricho, Fox, & Hameed, 2020).

Epoxies are a diverse group of heat-curing polymers that are utilized as coatings in the PIPS (Polymer Infiltrated Photonic Structures) process. Epoxy resins are primarily studied for their regulatory compounds due to the abundance of commercially available resin saps and curing agents. These resins and curing agents are combined to form copolymers that exhibit specific characteristics, such as a particular refractive index. In the PIPS method, the chemical curing process can either involve a chain reaction or a phase-separation reaction. The phase-separation reaction is particularly effective because it produces an optimal morphology of liquid crystal droplets dispersed within a continuous polymer matrix. This contributes to the desired optical properties and overall performance of the final product.

Two main concerns that influence drop diameter and thickness within the PIPS cycle are the types and relative convergence of the gear used, as well as the specific temperature. A fixed temperature significantly impacts the rate of chemical change, the thickness of the polymers, the dispersion rate of the liquid crystals, and the solubility of the liquid crystals within the polymers. Maintaining optimal temperature conditions is crucial for achieving the desired morphological and optical properties in the final PIPS product (Kashima et al., 2010). Each element is affected differently by specific temperatures, leading to significant variations in outcomes. Therefore, it is essential to carefully determine the optimal fixed temperature for each formulation. This consideration ensures that the desired results, in terms of both the morphological and optical properties of the final product, can be achieved effectively.

1.5 Epoxy Resin

Epoxy resins represent a broad category of flexible to rigid materials, highly sought after for their ability to be tailored for specific applications. The need for precise control over the properties of each formulation is essential, particularly in terms of the resin's processing capabilities. Epoxy materials are stable under ambient conditions, resulting in an extended shelf life for the resin and its epoxidized oligomers. Furthermore, epoxy formulations can react with a variety of agents. The choice of curing agent allows epoxy resins to be cured under ambient conditions, at elevated temperatures, or with exposure to light, resulting in durable, void-free products.

The field of epoxy resins has witnessed significant advancements, particularly in the realm of tailored formulations that enhance processing and application properties. The ability to fine-tune both resin and curing agent properties provides a tremendous opportunity for extensive research and commercial usage. This overview will focus on the vast array of epoxide materials developed since the 1930s, emphasizing commercially viable resins, curing agents, and supplementary additives. The evolution of these materials has been pivotal in meeting diverse industrial needs.

The recent decade has seen the introduction of novel gums and innovative drivers within the epoxy resin field. Several comprehensive reviews on epoxy materials will be referenced for a more in-depth examination of resin research. The field of Organic Compound Resins Science and Technology covers a wide range of operations, focusing on significant investigations related to the chemical reactions and curing processes of resins. A careful study of the resulting material properties and applications is crucial for understanding their industrial relevance. This ongoing research aims to enhance the performance and adaptability of these materials to meet an expanding array of applications (Tanaka, 2018).

The term 'resin tar' refers to a diverse set of In (Fig. 1-2. A), with chiefs (optic hub) of drops randomly positioned. Due to the refractive index difference between the compound and the fluid, violet light is dispersed upon incidence at PDLC oligomeric and compound materials. Whereas 'epoxy' alludes to an enormous ring having chemical element, the substances termed epoxy tars economically have express one, epoxides (three-membered oxirane rings) (Fig.1). Glycidyl will be used to name mixtures with epoxy for the gatherings. Even if organizations formed under current fix systems lack the epoxide gathering, they are still referred to as epoxy organizations.

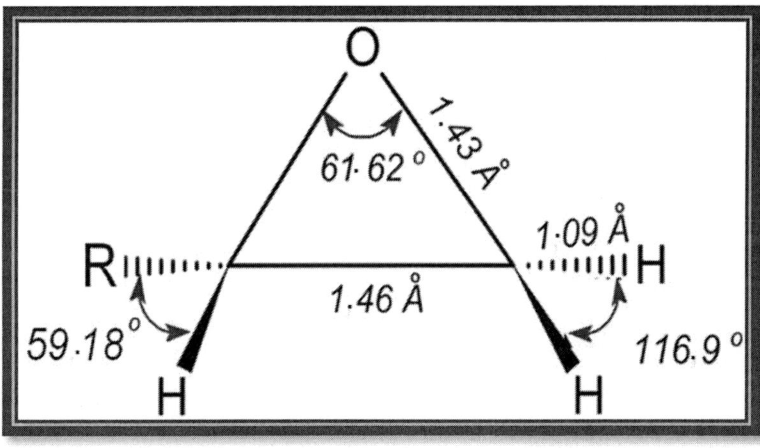

Figure 1-1 Oxirane ring found in common epoxy resins with bond angle and bond distance

1.5.1 Heat Curing Agents and epoxy monomers

Relieving specialists or thickeners are synthetic agents that help epoxy form cross-linked polymers, improving its unique properties. The most commonly used alkene series curing agents react with the oxirane ring along with active aliphatic compounds. The basic aliphatic series can be classified by their utility as primary (with two active hydrogens), secondary (with one active hydrogen), and tertiary (with no active hydrogens). Within the group of aliphatic curing agents, popular options include mercaptans, aliphatic amines, and polyamides. These agents play a significant role in enhancing the performance of epoxy resins in various applications (Vaz &

Montgomery Jr, 1987). Each has its properties, as shown in the below. They are:

A. Tri-ethylene-tetra mine (TETA)

It is used as a cross-linker an important class of epoxy hardener in the cured polymer network. It has two primary amines and two secondary amines. It's a straight amine construction and its refractive index is 1.496.

B. Diethylene-tetramine (DETA)

Diethylene-triamine is the next lowest molecular weight product. Clear and colourless with an ammonia-like odour, DETA is a one-component product containing two primary and one secondary nitrogen. The refractive index is 1.484.

C.Pentaerythritol-tetrakis-3-mercapto propionate (PERTMP)

It is another mercaptan/thiol chain hardener. It had four thiol (-SH) groups attached uniformly. Harder, which is not much more ideal than the amine group (NH2) in the formation of PDLC films.

D. Ethylene glycol diglycidyl ether (EGDE)

It is a bifunctional epoxy monomer. It has four substituted oxygen groups in the structure. In this kind of crosslinking agent, the opening of the epoxide ring must occur simultaneously with the occurrence of the crosslinking reaction, so it is used as a crosslinking agent for opening the epoxide ring and chemical interaction. It has a refractive index of 1.463

E. Bispheno A diglycidyl ether (BADGE)

Bisphenol A is a di-functional heat-curable epoxy monomer, also known as BPA it most commonly serves as a basic building block for polycarbonate and epoxy resin because of its unique properties. Its refractive index is 1.596

F. Neo pentyl glycol di glycidyl ether (NGDE)

Neopentyl glycol glycidyl ether is a penta-functional heat-curable epoxy molecule with two epoxy pendant groups at the end of the chain.it is majorly used as reactive diluent for epoxy resin. Its refractive index is 1.457

G. Nematic Liquid Crystal (NLC)

In a nematic phase, particles tend to align parallel to each other and with the director. The director is a unit vector that gives the preferred average molecular orientation about any point. The 4-Cyano-4'-n-pentylbiphenyl has been variously used. The refractive index value is n20/D 1.532.

1.6 PDLC and applications

The PDLCs are parallel fluid crystal bead scatterings in a constant polymeric framework, drawing a legitimate concern for analysts somewhat recently from both a fundamental and handy perspective. PDLCs are framed by spinodal deterioration instigated by warm extinguishing or polymer formation. These substances have superb magneto-optical character and extraordinary affinity in applications that need productive light dissipating. Now business usage incorporates switchable windows for protection control and enormous scope boards. The optical properties depend upon the world size, shape, and point request, not set in stone throughout the event stage, and reorientation components of the fluid semitransparent atoms restricted to the beads that happen throughout item use. PDLCs are fluid gem drops that are scattered haphazardly in a very complex network (Fig.1-2).

The typical fixation for polymers lies between 30%-50%. At the purpose field is applied at the presentation, the chiefs can organize themselves on course of applied field and light-weight is distributed through with insignificant dissipating (Fig. 1-2). As the purpose field is applied at the presentation, the chiefs organize themselves on the course of the applied field and light-weight is distributed through with insignificant dissipation (Fig. 1-2). The imagined polymers need to file of refraction close to the traditional refractive record (no) of fluid gems thus murkiness at the on-state may be forestalled. Because of their highly lightweight dissipating nature, monumental survey purpose, and splendor, no polarizers' area unit is needed for PDLC show activity. The numerous uses of PDLCs are switchable windows. The foremost well-known assortment of PDLC is the optically dissipating structure.

The convergence of compounds within fluid jewelry typically occurs within a rough timeframe of around half an hour. This process enhances the compound when integrated into LC/polymer emulsions, leading to the separation of fluid gems within the polymer's structure. Beads, which are a common feature, are usually of small size on the scale (Bulgakova et al., 2008). Liquid gem particles inside every drop have a limited request, notwithstanding, every bead can be haphazardly adjusted compared with others. The mix of drop size and isotropic direction of drops prompts a profoundly optically dissipating state and gives the phone a smooth appearance. At the point when a similar material is then exposed to an electric field, electro-optic reorientation of the fluid precious stone beads happens. This then, at that point, lessens the level of optical dissipating through the cell, bringing about a straightforward

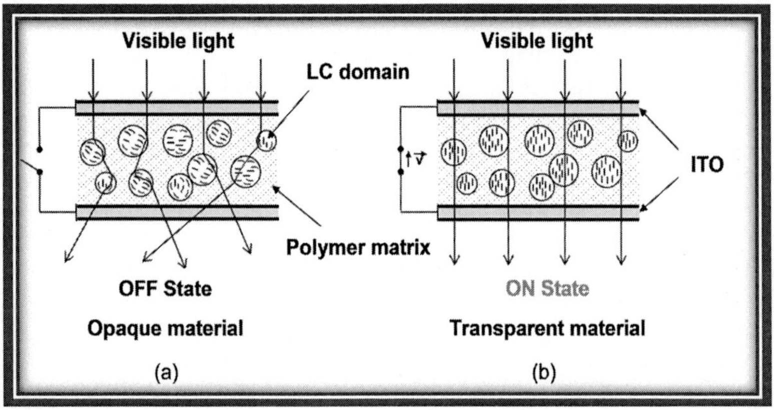

Figure 1-2 Diagram of a polymer-dispersed liquid crystal (PDLC) a.) In the off-state where light is scattered and b.) In the on-state where light is transmitted through the display

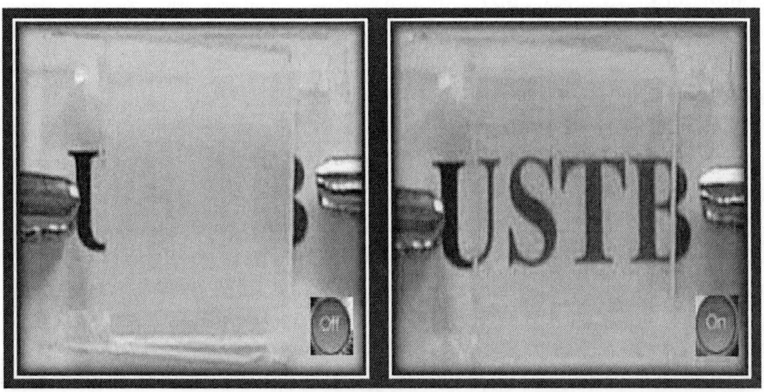

Figure 1-3 Schematic drawing of the PDLC film electro-optical application

Dispersing PDLCs utilizing micro-sized drops, for example, has varied usages. Switchable windows/security screens (changeable between a straightforward/clear state and a misty/dissipating state) are currently monetarily accessible and use PDLC innovation. Artificial colours will likewise be incorporated in PDLC combos so that they often particularly dissipate inexperienced, red, or blue lightweight on an individual basis. Besides, the rising bead space region (in between the fluid gem on the polymers) brings solid securing energies and therefore quick exchanging times that are of extra advantage to the presentation business.

Implementing PDLCs is challenging due to their higher voltage requirement of 0.2 to 10 V/micron, compared to 0.1 to 5 V/micron for traditional liquid crystals. This can complicate the design and integration of PDLC systems.It prompts increased power utilization and far more expensive running hardware. However, as a result of impulsive arrangement of the 'off' specific, full fluid jewellery birefringences cannot be used, with gadget transfer occurring with a diminished stroke. All the same, this can be to some extent remunerated by utilizing thicker gadgets, because their exchanging time is generally free of cell thickness as displayed in Fig.4.

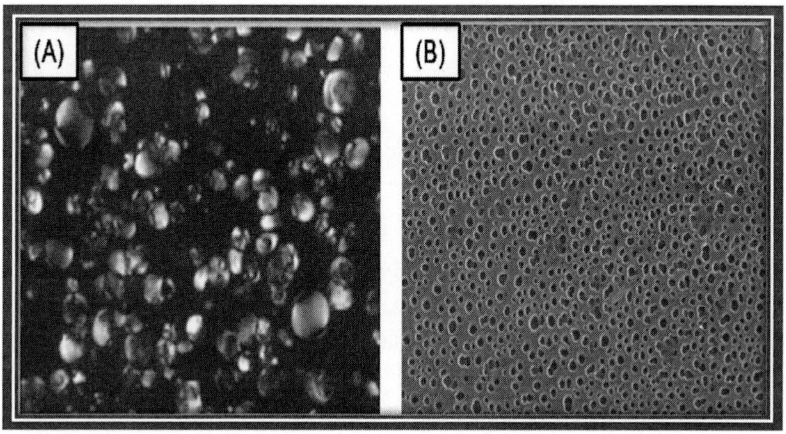

Figure 1-4 (A) A Polarizing microscope image of the PDLC film and **(B)** Scanning electron microscope photograph of a cross-section of a PDLC film.

The setup of the liquid crystal beads in a polymer grid is the focal point of much momentum research. A wide range of arrangements was noticed and they rely upon variables, for example, bead thickness and structure, surface mooring, and applied fields. This part will portray the absolute more normal designs. Schematic delineation of nematic chief arrangements in various round pits, we see in (Fig 5).

For a circular drop, no versatile contortion is required to align a bipolar bead with an electric field. However, the drops in PDLC (Polymer Dispersed Liquid Crystal) substances are often not perfectly circular. The random orientation of bipolar beads in a PDLC film arises from the distribution of shapes and orientations of somewhat elongated drops. By analyzing this variation in drop structure, it is possible to calculate the influence of drop morphology on the driving voltage of a PDLC display that incorporates bipolar beads. This understanding can help optimize the performance of such materials in various applications (McFarland, Koenig, & West, 1993).

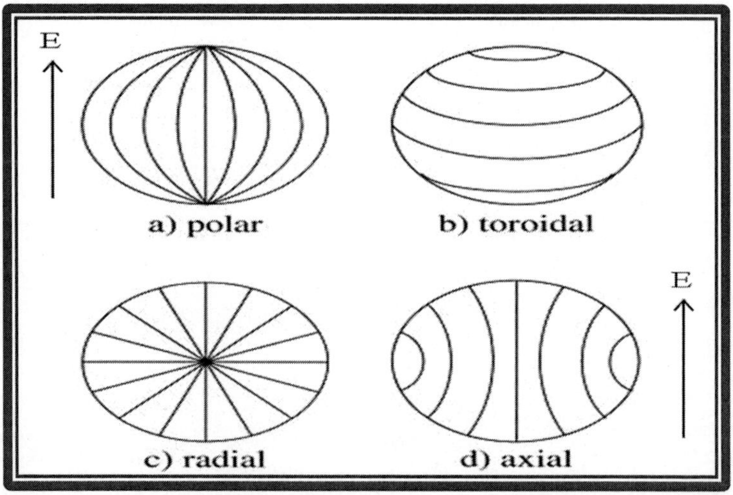

Figure 1-5 Schematic illustration of nematic director configurations in spherical cavities: (a) the bipolar configuration aligned by an electric field,(b) the toroidal configuration (c) the radial configuration, (d) the axial configuration in an applied field

1.7. Discovery of Liquid Crystals: The Intermediate Phase

It was taught in colleges that 3 periods of issue: solids, fluids, and gases. Stages controlled through atoms area unit settled requested regarding each other. The Temperature vacillation builds atoms to vibrate or move which will prompt touch on move between varied phases. Particles in solids area unit mounted and should be exceptionally requested and settled in three-layered clusters (translucent solids) and discretional places (nebulous solids).

In isotropic fluids, such as gases, particles move freely without a fixed orientation or point of reference. This means that the particles are randomly distributed and exhibit no preferred direction for their alignment. As a result, their behaviour is governed solely by thermal motion, leading to a uniform distribution of velocities and directions. This characteristic is a key feature of isotropic media, influencing how these fluids interact with external forces, such as electric or magnetic fields. The distinction between the two states is that the atoms of gases are so much separated from each other, which makes gases compressible. In contrast, the particles in fluids have an attractive force that keeps them close to each other, making fluids non-compressible. As the logical instruments and innovation progressed, therefore did our insight concerning completely different periods of a problem aside from solids, fluids, and gases. In 1888, associate degree Friedrich Reiniztcr Austrian biologist was warming substances unremarkably occurring in plants on an associate degree shoot underneath a polarizing magnifying lens.

This substance, powerful at 25°C degrees, was heated into associate degree overcast fluid at 145.5°C degrees; because the substance was warm at 178.5°C degrees, the fluid was clad to be clear. He talked about his perception along with his partner Otto Lehmann, a German scientist. Lehmann begat this revelation as a "fluid gem" since it showed each the atomic request of glasslike solids and the progression of the same sort of fluids (Collings, 2002). This substance is known as cholesteryl benzoate (C34H50O2). The structure is shown in (fig. 6).

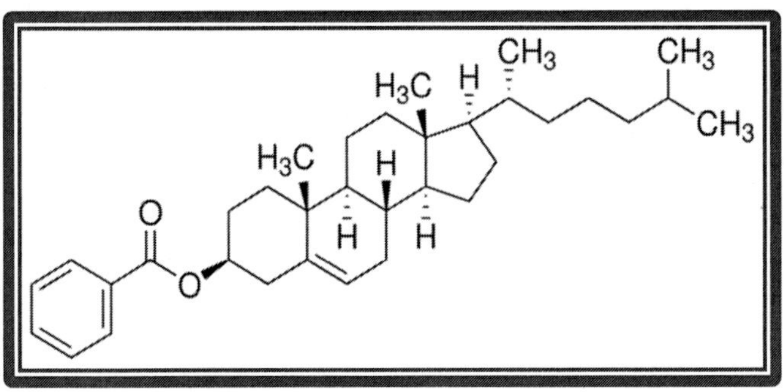

Figure 1-6 Cholesteryl benzoate ($C_{34}H_{50}O_2$).

Liquid crystals can be found in nature as well as synthesized for various applications. Their structures are commonly categorized into two types: calamitic and discotic. Calamitic liquid crystals are typically characterized by an elongated, rigid aromatic core paired with a flexible organic tail. These rod-like molecules tend to align parallel to one another, which gives them their unique properties. In contrast, discotic liquid crystals are shaped differently and can exhibit different behaviours due to their disk-like structure. This distinction between calamitic and discotic forms is essential for understanding their applications in displays and other technologies. Discotic liquid crystals contain hexagonal rings at the centre and these structures tend to stack on high of each other. The additional types of liquid crystals, such as curve-core (or banana) shaped liquid crystals and chemical compound liquid crystals, also undergo transitions depending on their structure and conditions. Thermotropic liquid crystals transition between phases based on temperature variations. For instance, typical thermotropic liquids have a molecular structure that combines both a rigid component, such as benzene or aromatic rings, with a more flexible hydrocarbon arrangement, allowing for notable elasticity. On the other hand, lyotropic liquid crystals are typically amphiphilic, possessing hydrophilic polar head groups and hydrophobic organic tails. Their phase behaviour is influenced by the concentration of the surrounding medium, such as water or other solvents, leading to distinct liquid crystalline phases based on the interactions between these components and the solvent. Understanding these variations helps in optimizing the properties and applications of liquid crystal materials (Dierking, 2009).

The thermotropic calamatic fluid is a crystal modification in between the semi-transparent fluid and therefore the isotropic fluid through mesophases. Fluid gem mesophases are often recognized by setting a fluid gem take a look at between 2 crossed polarizers beneath a magnifying instrument. The atoms within the nematic stage, displayed in (Fig.7), are organized typically one next to the other, having long-range orientation order. Nonetheless, there is no point order (Chandrasekhar, 1978).The approximate leading way of those molecules is understood because of the director which is denoted because of the vector $\vec{n}$.

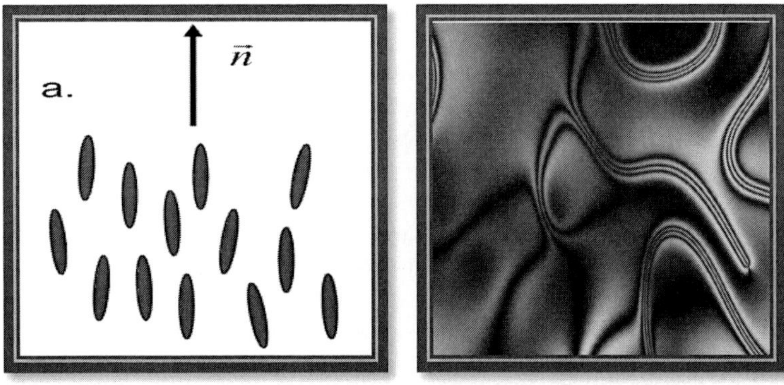

Figure 1-7 Nematic phase exhibited by a.) A schematic representation of liquid crystal molecules in which the molecular director shows the average direction the molecules are oriented; b.) A photographic image of a hybrid aligned nematic phase

1.8 Introduction to Liquid Crystal

More than a single transition between the solid and liquid phases occurs in many organic materials. These materials have one or more intermediate phases. These phases, called mesophases, have both solid-like molecular order and liquid-like fluidity. Mesophases possess a crystal lattice structure, but no rotational order is known as disordered crystal mesophases. Mesophases having this kind of order crystal are known as liquid crystals and are generally composed of monomers shows the chemical formulas of the following typical rod-like liquid crystals: 4-pentyl-4'cyano biphenyl (5CB).

The compound 5CB is often the primary component in liquid crystal mixtures used for PDLC fabrication. Current research is concentrated on thin-film materials, with a strong emphasis on the development of functional liquid crystal materials. These include super twist LCDs, curved and twisted nematic LCDs, ferroelectric LCDs, and bistable LCDs. Among the innovations in liquid crystal technology, polymer dispersed materials (PDLC) have proven to be particularly effective and are making significant strides in the display industry (Huihui Wang et al., 2015).

Classification of Liquid Crystal

Phases based on their molecular order is presented as follows:

1.8.1 Nematic Order

In a nematic phase, particles tend to align parallel to each other and with the director, which is a unit vector representing the preferred average molecular orientation at any given point. The nematic phase is the simplest naturally occurring liquid crystal (LC) phase, characterized by rod-shaped LC molecules (mesogens) that exhibit long-range orientational order, aligning their longer axes approximately parallel to one another while lacking long-range translational order. These nematic liquid crystals can reorient, and the electro-optical properties of any polymer-dispersed liquid crystal (PDLC) film are influenced by the specific nematic liquid crystal used.

1.8.2 Cholesteric Order

Cholesteric order closely resembles nematic order at a local level, with molecules aligning along the director. Cholesteric phases arise from chiral nematic molecules, making "cholesteric" and "chiral nematic" effectively synonymous. In polymer-stabilized cholesteric liquid crystals (PSCL), the transparent state features an in-plane texture. However, when there's a discrepancy in the refractive indices between the polymer network and the liquid crystal, the material becomes opaque without an electric field. When an electric field is applied, the cholesteric liquid crystal aligns in a planar texture perpendicular to the liquid crystal cells, resulting in high transparency (Zhao et al., 2021).

1.8.3 Smectic Order

The smectic phase exhibits one degree of translational order, resulting in a layered structure. There are at least ten identified smectic phases, with smectic phase a being the most well-known. In smectic-A ordering, the molecules align parallel to the sheet normal within each layer. However, if the particles are uniformly bent away from the sheet normal in each layer, the ordering changes, leading to different smectic structures. This bending can significantly affect the properties of the liquid crystal, influencing its behaviour under various conditions.

1.9 The Order of Liquid Crystals (LCs)

The liquid crystal chief is uniform within the nanometer range but shifts during fabric processing. The LC. However, the sub-atomic gatherings square measure requested could be a basic property of the fluid gem, and sub-atomic requests are communicated by the related condition (Equation 2-1).

$$S = \frac{1}{2}(3\langle Cos^2\theta\rangle - 1) \qquad (2\text{-}1)$$

The request boundary is indicated by S, and θ is the place where the particle area unit is set regarding the chief. Condition 2-1 may be a second-request Legendre polynomial. The sections within the state of affairs show that since θ for each atom vacillates due to heat dangers, the points of direction ought to have got wind of the centre of out. In a wholly irregular direction S=0; nevertheless for superb arrangement, S=1. In fluid jewellery stages, S will go from zero 3 to 0.9 (Collings, Fisch, & Mooney, 1992).

Yet, because the temperature is swollen to the isotropic stage, the request boundary diminishes. In (Fig.1-8), a schematic portrayal of the request boundary of nematic fluid jewellery is thought for temperature. The request boundary variables are in the nematic stage. When the boundary reaches zero, the isotropic stage is achieved (TN1), marking the transition from nematic to isotropic in the first request.

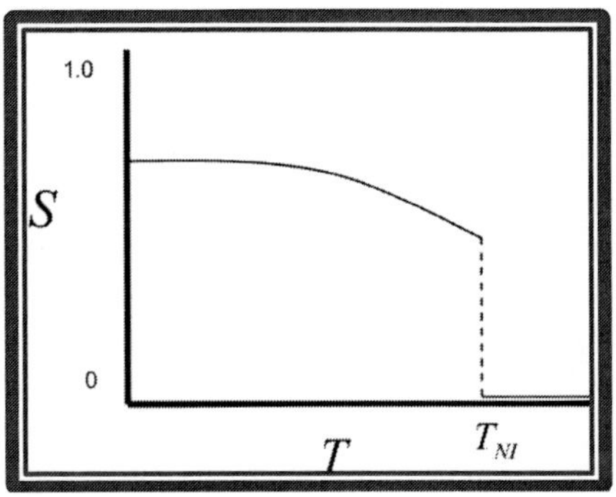

Figure 1-8 A plot of the order parameter S and its dependence on temperature for a typical nematic liquid crystal

1.10 Deformations of Liquid Crystals

In wonderful arrangement conditions, LC atoms are organized next to each other all through the medium; notwithstanding, even with the littlest outside powers, for example, electromagnetic or mechanical, the chief can without much of a stretch be twisted. The chief can be disfigured into numerous setups however they can be a blend of or one of these three essential distortions: spread, wind, and twist (Fig. 9). This spread, turn, and curve disfigurements can be signified as the accompanying regarding flexible constants K11, K22, and K33.

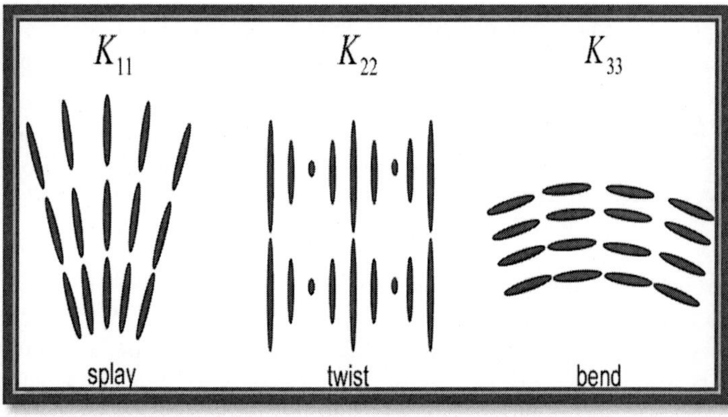

Figure 1-9 Three basic types of deformations: splay, twist, and bend

The upsides of the versatile constants typically appear around 10^{-12} N, indicating that as the versatile constant decreases, the tendency for the fluid crystal manager to twist increases. The versatile constants for a fluid crystal are still determined tentatively by applying external electrical or magnetic fields to an LC between materials with treated substances.

1.11 Optics of Liquid Crystals

"Liquid crystals, owing to their properties, exhibit optic refraction." Double refraction is an interesting phenomenon that occurs when the speed of light in a material differs depending on its polarization. When light passes through a fluid jewel medium, its refraction behaves differently due to the unique properties of the material. This process is known as double refraction, where the light is split into two distinct rays, each with a different refractive index based on its polarization. This unusual refractive behaviour is referred to as the extraordinary refractive index, contrasting with the typical refractive index that applies when light interacts with the fluid jewel material uniform, without considering polarization. The double refraction of the fabric is essentially the excellence between the two refractive records shown in (Fig. 1-10, 11) (Hecht & Zajac, 2002).

$$Dn = n_{//} - n_{\wedge} = n_e - n_o \quad (2\text{-}2)$$

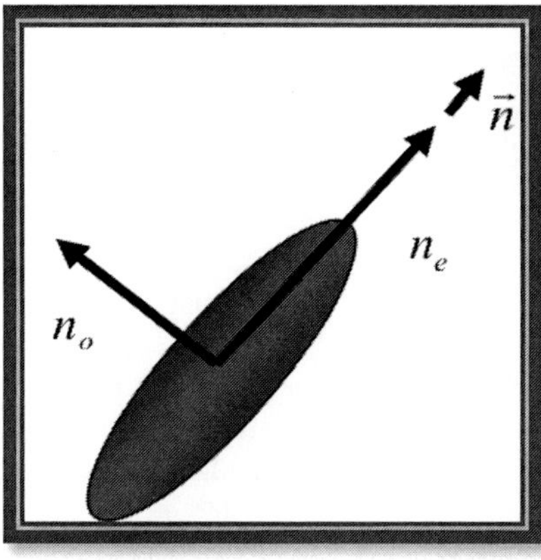

Figure 1-10 Schematic representation of a liquid crystal molecule and its refractive indices

The birefringence depends on the temperature in liquid crystals (Hicks, 2006). At the point when the LC is warmed, the birefringence consistently diminishes until the isotropic stage is reached. From that point onward, just the refractive record of the isotropic fluid is available (Fig.11).

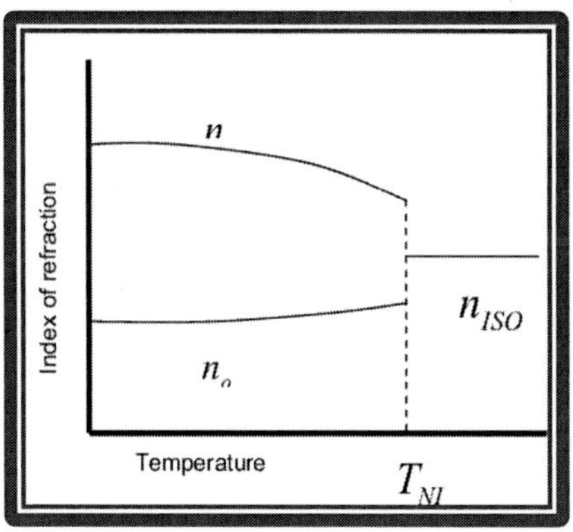

Figure 1-11 Graph of the index of refraction for a typical nematic liquid crystal over temperature.

An optical phenomenon that typically occurs with liquid crystals is light scattering. Light scattering occurs in two main ways: one involves reflection from very rough surfaces where the angles of light reflection are random, and the other results from inhomogeneities within the material. These inconsistencies can cause variations in how light is transmitted or reflected, leading to unique scattering patterns. This behaviour is crucial in determining the optical properties of liquid crystals and can impact their applications in displays and other technologies (Hulst & van de Hulst, 1981). For the instance of fluid gems, being anisotropic, light-weight dissipating is traditional.

Once the fabric is produced in bulk, the chief alignment of the liquid crystal can vary throughout its volume. As light passes through the liquid crystals, it interacts with the fluid crystal particles, causing them to vibrate in response to the optical input and scatter the light at irregular angles. As a result, this scattering effect can make the liquid crystals appear cloudy or shaded. This phenomenon is essential to understanding how liquid crystals function in various applications, particularly in display technology.

1.12 Electrical Properties of Liquid Crystals

Fluid precious stones possess charges that circulate along the long axis of each atom. In the presence of electric fields, these charges can either separate or move in response to the applied field. When the electric field is strong enough, it can cause the atoms to rotate. Fluid crystals are dielectrically anisotropic, which means they have two distinct permittivities, the parallel permittivity ($\varepsilon//$) and the perpendicular permittivity ($\varepsilon\perp$). The dielectric property describes the difference between these two permittivity, which plays a crucial role in their behaviour within electric fields and their applications in various technologies (Hicks, 2006).

$$\Delta\varepsilon=\varepsilon_{//}-\varepsilon_{\perp} \tag{2-3}$$

On the off probability that $\Delta\varepsilon >$ zero, the fluid gem chief can usually regulate adore the applied electrical field; otherwise ($\delta\varepsilon < 0$), it'll have the propensity to regulate opposite to the electrical field (Fig.1-12).

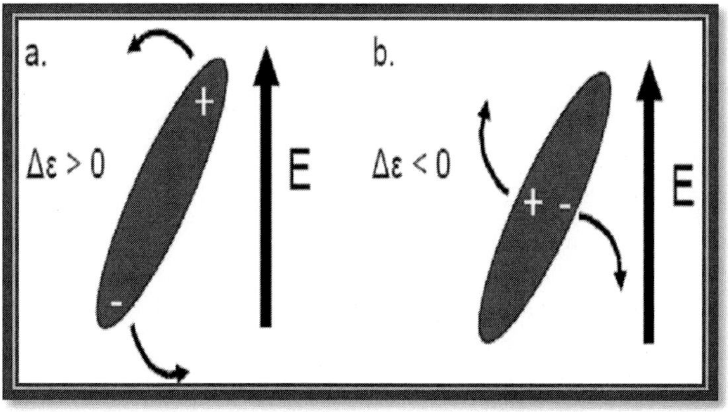

Figure 1-12 Schematic representations of liquid crystal molecules rotating in response to external electric fields when a.) $\Delta\varepsilon > 0$, where the molecule aligns parallel to the field and b.)$\Delta\varepsilon < 0$ where the molecule aligns perpendicular to the field.

1.13 Applications of Liquid Crystals

Fluid precious stone innovation has had a significant impact in numerous spaces of science and design, just as gadget innovation. Applications for this extraordinary sort of material are yet to be found and keep on giving effective answers for some different issues.

1.13.1 Liquid Crystal Displays (LCD)

The most widely recognized use of fluid crystal technology is fluid crystal displays (LCDs). LCDs take advantage of the properties of liquid crystals, enabling them to manipulate light and create images on screens. This technology is commonly found in televisions, computer monitors, and various portable devices, providing high-quality visuals while being energy-efficient. The intricate behaviors of liquid crystals, influenced by external electric or magnetic fields, contribute to the display's functionality and clarity. This field has developed into a multi-billion-dollar industry and numerous significant scientific and design revelations have been made. The standard of LCD activity is displayed in the following (Fig.1-13).

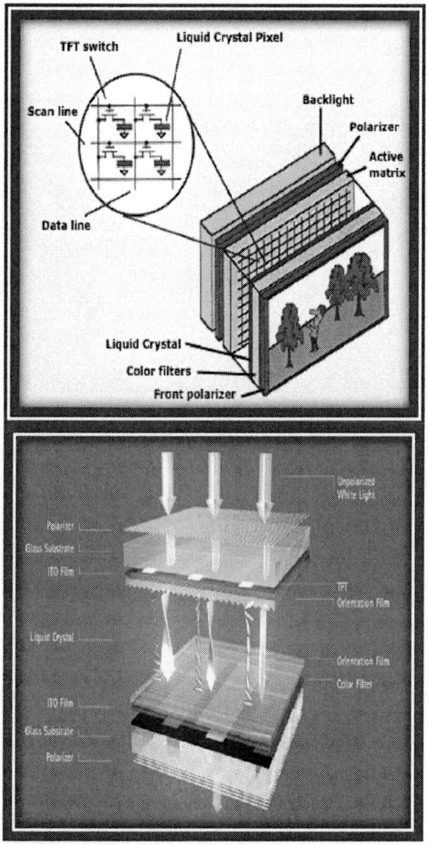

Figure 1-13 Active-matrix liquid crystal display

1.13.2 Optical Imaging

The use of fluid gems that are just being investigated is optical imaging and recording. During this innovation, a fluid gem cell is placed between two layers of photograph conductors. Lightweight is applied to the photograph conductor, building the material's conduction. This causes an electrical and electrical produced within the fluid gem compared to the facility of the sunshine. The electrical example will communicate by associate anode, empowering the image to be recorded.

1.13.3 Polymer dispersed liquid crystals

PDLCs comprise fluid gem beads that were area unit scattered during a compound network study (see Fig.14). The following material could be a quiet "Swiss cheddar" compound

with fluid gem beads filling within the openings. These minuscule beads (several microns across for right down-to-earth applications) are unit accountable for the one-of-a-kind conduct of the fabric. By ever-changing the direction of the fluid gem particles with an electrical field, it's possible to fluctuate the force of communication (K. Li et al., 2020).

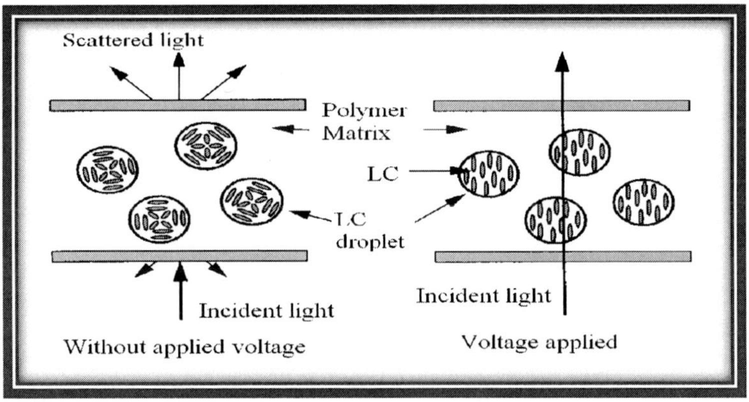

Figure 1-14 In a typical PDLC sample, there are many droplets with different configurations and orientations. When an electric field is applied, however, the molecules within the droplets align along the field and have corresponding optical properties

1.13.4 Other Liquid Crystal Applications

This strategy is utilized for the perception of RF (radio recurrence) waves in waveguides. They're utilized in clinical applications wherever, for example, transient strain communicated by a mobile foot on the bottom is calculable. Low molar mass fluid precious stones have applications as well as eradicable optical circles, full-tone "electronic slides" for laptop-supported drawing, and lightweight modulators for shading electronic imaging.

1.14. Electro-optical materials and their properties

Materials that alternate their optical characteristics on electrical fields are known as electro-optical materials. The field of electro-optical materials is by no means new. Since that time, electro-optical effects in some liquids and crystals have become known. First reported in 1985 Polymer dispersed liquid crystals elevated the type of electro-optical substances to numerous capacity programs creating extensive medical and business interest. The common methods of preparation are three-phase separation techniques; the separation phase is either through polymerization, temperature, or solvent evaporation. The light absorption in these materials can be considered negligible. The change in transmittance means the change in the light scattering. PDLC is a subclass of liquid crystal dispersion films, composed of organic polymer matrix and liquid crystal (LC).

Microdroplets in PDLC (polymer dispersed liquid crystal) films present significant challenges for the effective operation of these materials across various applications. Much research has focused on reducing the driving voltage needed for these films to function properly. However, there are still very few PDLC materials that can simultaneously meet the three critical performance indicators: low driving voltage, high contrast ratio, and minimal lag time. To enhance the overall performance of PDLC materials, further investigation and development are required. The goal is to create PDLC thin-film materials that boast exceptional electro-optical properties. Achieving a significant reduction in response time while maintaining a high contrast ratio and low driving voltage remains a key objective in the advancement of PDLC technology (Kizhakidathazhath, Nishikawa, Okumura, Higuchi, & Kikuchi, 2020).

1.15 Factors Affecting on Electro-optical Performance

The properties of PDLC films can vary significantly due to the different substances used, the bead shape, and the film development process. With the advent of transparent window technologies, various solutions have emerged to tackle the complex challenge of providing compensatory protection while allowing for the free flow of light. Traditional solutions like blinds, curtains, and screens often restricted visibility, telling onlookers to "stay out of other people's affairs." However, with the introduction of switchable light protection glass, these previous limitations have been overcome. Now, you can enjoy ample natural light while having the option for complete privacy at the flick of a switch, achieving the perfect balance between visibility and protection (J. W. Doane, 1990).

The concept revolves around a fluid gem sheet sandwiched between two layers of glass, which remains transparent when powered on. In this state, the glass allows light to pass through seamlessly. However, when the power is turned off, the glass transforms quickly, becoming opaque. This transformation is attributed to the behaviour of the liquid crystal molecules within the layers. When no voltage is applied, the liquid crystal molecules are randomly oriented, leading to greater light scattering and thus a foggy appearance. On the other hand, when voltage is applied, these molecules align in a specific arrangement that permits light to travel through the glass uniformly, allowing for a clear view. This innovative technology creates a versatile solution for situations requiring privacy or light control without sacrificing natural illumination (Montgomery Jr, West, & Tamura-Lis, 1991).

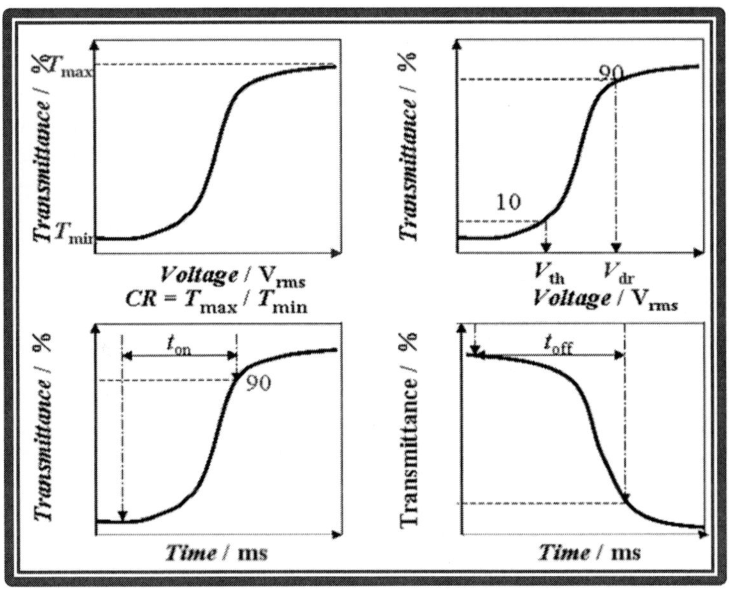

Figure 1-15 A schematic diagram of the transmittance applied voltage curves of PDLC films

Positive characteristics kept in mind above are clearness and transfer of the film for the off and on states, less driving voltage, less power utilization, high exchanging times, and highest film obstruction as displayed in (Fig.15). Since these characteristics are connected, it is ordinarily impractical to remodel them autonomously.

1.15.1 Driving Voltage

In the setup of switchable glass, anodes are attached to simple cathodes. When no voltage is applied, the liquid crystals are randomly arranged, which leads to a diffuse scattering of light that gives the glass a "smooth white" appearance. This scattering effect obscures visibility. When a voltage is applied to the anodes, an electric field is created between the two cathodes on the glass. This field causes the liquid crystals to align, allowing light to pass through the droplets with minimal scattering and resulting in a clear state. Most modern switchable glass products operate primarily in either an "on" or "off" mode, providing a straightforward method for controlling visibility and privacy (Firehammer, Crawford, & Lawandy, 1998).

In any case, the capacity is there to make a dynamic level of straightforwardness to force by the applying voltage. It is often conceivable as a result of, at lower voltages, a few of the fluid gems regulate whole within the force field, therefore simply a touching piece of the sunshine goes through whereas most of the sunshine is distributed. Because the voltage is distended, less fluid gems keep askew, delivery concerning less light-weight being distributed. The Vth and Vsat square measures are characterized by the electrical fields required for the conveyance to achieve 100% and 90% on an individual basis (Petti, Mormile, & Blau, 2003). It is by and hugely conceded that the dimensions of the LC drop encompass a solid relationship with electro-optical execution in an exceedingly PDLC framework. In keeping with as so much as anyone is bothered, the Vth may be a direct capability of associate degree equal structure of the beads R as per,

$$V_{th} = \frac{d}{R}\left[\frac{K(\omega^2 - 1)}{\varepsilon_0 \Delta\varepsilon}\right]^{\frac{1}{2}}$$

(3-1)

$$A = \frac{V_{sat}\sqrt{\varepsilon_0 \Delta\varepsilon K}}{d}\tanh\left[\frac{V_{sat}}{2}\sqrt{\frac{\varepsilon_0 \Delta\varepsilon}{K}}\right] \tag{3-2}$$

Where d, K, ω, εo and Δε address film thickness, versatile steady, angle proportion, Vacuum stuff consistent, and stuff property of the LC singly, For the foremost half, the sting voltage V_{th} increments with decreasing LC drop size; incoming regarding immersion Voltage V_{sat} changes at an equivalent time.

1.15.2 Response Time

Contrast proportion (CR) could be an important proportion of the electro-optical properties in an exceedingly PDLC framework. The contrast ratio of PDLC films is employed to portray the distinction between an easy state and an obscure state characterized as, When the film is in its transparent state, it allows light to pass through easily, creating a clear view. Conversely, in its opaque state, the film scatters light, resulting in a frosted appearance that obscures visibility.

$$Cr = \tau r/\tau d \tag{3-3}$$

Where wherever and τd are a definitive ascent time conveyance and also the underlying rot time conveyance separately. The ascent time (τr) and decay time (τd) are determined through time reliance of the transmission constant. Rise time is processed because the time require to travel from 100 % to 90 % of the foremost extreme Conveyance of the instance when turn on and that is essential to travel From 90% to 100% of the best conveyance of check winding down thought as rot time.

$$t_{on} = \frac{1}{\gamma_1}\left[\frac{9\varepsilon_0 \Delta_\varepsilon V^2}{d^2\left(\frac{\rho_p}{\rho_{lc}}+2\right)^2} + \frac{K\left(l^2-1\right)}{a^2}\right] \tag{3-4}$$

$$t_{off} = \frac{\gamma_1 \times a^2}{K\left(l^2 - 1\right)} \tag{3-5}$$

Where K, εo, Δε, d, ρp, ρlc, l and Ɣ versatile consistent of fluid gem, Vacuum permittivity, insulator property of fluid jewel, the thickness of PDLC film, resistance of the chemical compound, resistance of the fluid gem, the proportion length of the semi-significant pivot and constant of consistency of the fluid gem substances on an individual basis.

1.15.3 PDLC Innovations

PDLC film is a little while ago arriving at commercialization. A few applications, like gathering rooms, business structures, adaptable showcases, switchable windows, greenhouses, and other innovative presentation gadgets as displayed (Kato, Chen, & Pac, 2004).Late developments in PDLC films are expanding the potential applications for these materials and improving their overall performance. This progress is crucial for enhancing the functionality and versatility of PDLC technology in various fields.

Figure1-16 PDLC in the application of electronic intelligent glass ON and OFF State

1.16 Silver Nanoparticles (AgNPs)

The prefix nano within the word engineering science implies a billionth of the (1x10-9). The engineering science manages different constructions of problems having aspects of the request for a billionth of a meter. Plasmonic metal nanoparticles, like gold, silver, and platinum, exhibit effective light-engrossing and dissipating properties, the explanation of such properties is expected to the plasmon reverberation. Frequencies of light can cooperate with the conduction electrons in the plasmonic metal nanoparticles thus the electrons will waver, all things considered, a peculiarity known as surface plasmon reverberation (SPR).

These invigorated resonances can significantly enhance the retention and dissipating forces compared to other non-plasmonic particles of similar size. Interestingly, the interactions between electrons in pure metals at a similar mass scale and light have more limitations, resulting in the absence of observed plasmon resonance (Gui, Zheng, Wang, Li, & Zhuang, 2015).

Both the dissipating and absorbance of light play crucial roles in the apparent shading of materials. The yellow tone observed in silver nanoparticles is likely a result of plasmon absorbance. A colloid refers to a system where tiny insoluble particles are consistently dispersed throughout another substance. In this experiment, a colloid is formed by uniformly scattering insoluble silver nanoparticles throughout the solution. This uniform distribution contributes to the unique optical properties of the colloid, enhancing its colouration and interactions with light (Liu et al., 2019).

Silver nanoparticles are stabilized through the adsorption of borohydride particles, which form a protective layer around each nanoparticle. The negatively charged borohydride particles create electrostatic repulsion between the silver nanoparticles, preventing them from aggregating. This stabilization allows the silver nanoparticles to remain dispersed in the solution, maintaining their unique optical properties and enhancing their interactions with light.

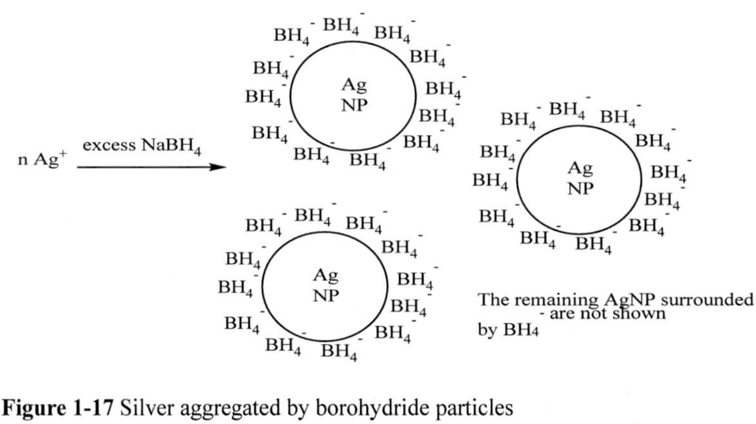

Figure 1-17 Silver aggregated by borohydride particles

References

Bulgakova, SA, Mashin, AI, Kazantseva, IA, Kashtanov, DE, Jones, MM, Tsepkov, GS, Nezhdanov, AV. (2008). Influence of the composition of the polymer matrix on the electrooptical properties of films with a dispersed liquid crystal. *Russian Journal of Applied Chemistry, 81*(8), 1446-1451.

Capricho, Jaworski C, Fox, Bronwyn, & Hameed, Nishar. (2020). Multifunctionality in epoxy resins. *Polymer Reviews, 60*(1), 1-41.

Castellón, Erick, & Levy, David. (2018). Smart Windows Based on Liquid Crystal Dispersions. *Transparent Conductive Materials: Materials, Synthesis, Characterization, Applications,Weinheim,Germany,5*(4), 37-365.

Chen, Tien-Jung, Chen, Yu-Fan, Sun, Chia-Hsing, & Wu, Jin-Jci. (2006). Electro-optical properties of reverse-mode films of planar aligned polymer-dispersed liquid crystal. *Journal of Polymer Research, 13*(2), 85-89.

Collings, Peter J. (2002). *Liquid crystals: nature's delicate phase of matter*: Princeton University Press.

Collings, Peter J, Fisch, Michael R, & Mooney, Mary A. (1992). Liquid crystals: nature's delicate phase of matter: American Association of Physics Teachers.

Doane, JW, Vaz, NA, Wu, B-G, & Žumer, Slobodan. (1986). Field controlled light scattering from nematic microdroplets. *Applied Physics Letters, 48*(4), 269-271.

Drzaic, Paul, & Drzaic, Paul S. (2006). Putting liquid crystal droplets to work: a short history of polymer dispersed liquid crystals. *Liquid crystals, 33*(11-12), 1281-1296.

Fergason, James L. (1985). Polymer encapsulated nematic liquid crystals for display and light control applications. *SID Dig. Tech. Paper, 16*, 68-70.

Hakemi, H, Khenkin, M, Tamari, R, & Gal-Fuss, D. (2020). Transient Phenomenon in Polymer-Dispersed Liquid Crystals. *Recent Progress in Materials, 2*(1), 1-1.

Hecht, Engene, & Zajac, A. (2002). Optics forth edition: Addison Wesley.

Higgins, Daniel A, Hall, Jeffrey E, & Xie, Aifang. (2005). Optical microscopy studies of dynamics within individual polymer-dispersed liquid crystal droplets. *Accounts of chemical research, 38*(2), 137-145.

Hui, LI, Feifan, YANG, Jing, PENG, Fan, PAN, Yuntao, WU, Zhang, Yanduo, & Xiaolin, XIE. (2021). Electro-Optical Properties of a

Transmissive Diffuser based on Electrically-Tuneable Polymer-Dispersed Liquid Crystal Film. *Materials Science, 27*(1), 8-12.

Hulst, Hendrik Christoffel, & van de Hulst, Hendrik C. (1981). *Light scattering by small particles*: Courier Corporation.*84*(360),198-199 .

Klapetek, Petr, Valtr, Miroslav, Nečas, David, Salyk, Ota, & Dzik, Petr. (2011). Atomic force microscopy analysis of nanoparticles in non-ideal conditions. *Nanoscale research letters, 6*(1), 1-9.

Li, Ke, Jiang, Haodong, Cheng, Ming, Li, Yifei, Yin, Zhen, Luo, Dan, . . . Liu, Yan Jun. (2020). Controlling morphological and electro-optical properties via the phase separation in polymer/liquid-crystal composite materials. *Liquid Crystals, 47*(2), 238-247.

Liu, Yourong, Zheng, Jihong, Shen, Tong, Wang, Kangni, & Zhuang, Songlin. (2019). Diffusion kinetics investigations of Nano Ag-doped holographic polymer dispersed liquid crystal gratings. *Liquid Crystals, 46*(12), 1852-1860.

McFarland, Coleen A, Koenig, Jack L, & West, John L. (1993). Influence of the polymer matrix on the bipolar and radial droplet configurations within PDLC films examined by infrared spectroscopy. *Applied spectroscopy, 47*(5), 598-605.

Mohan, Pragyan. (2013). A critical review: the modification, properties, and applications of epoxy resins. *Polymer-Plastics Technology and Engineering, 52*(2), 107-125.

Mormile, P, Musto, P, Petti, L, Ragosta, G, & Villano, P. (2000). Electro-optical properties of a PDLC based on unsaturated polyester resin. *Applied Physics B, 70*(2), 249-252.

Paluvai, Nagarjuna Reddy, Mohanty, Smita, & Nayak, SK. (2014). Synthesis and modifications of epoxy resins and their composites: a review. *Polymer-Plastics Technology and Engineering, 53*(16), 1723-1758.

Pandian, A Muthu Kumara, Karthikeyan, C, Rajasimman, M, & Dinesh, MG. (2015). Synthesis of silver nanoparticle and its application. *Ecotoxicology and environmental safety, 121*(10), 211-217.

Tanaka, Yoshio. (2018). Synthesis and characteristics of epoxides, *Epoxy resins: chemistry and technology* (2nd ed ., pp. 9-283).United kingdom, England:Routelege taylor and francis England.

Vaz, Nuno A, & Montgomery Jr, G Paul. (1987). Refractive indices of polymer-dispersed liquid-crystal film materials: Epoxy-based systems. *Journal of applied physics, 62*(8), 3161-3172.

Zhang, Menghua, Zheng, Jihong, Gui, Kun, Wang, Kangni, Guo, Caihong, Wei, Xiaopeng, & Zhuang, Songlin. (2013). Electro-optical characteristics of holographic polymer dispersed liquid crystal gratings doped with nanosilver. *Applied Optics, 52*(31), 7411-7418.

Zhao, Rui, Li, Xiaoshuai, Wang, Kemeng, Huai, He, Ma, Hongmei, & Sun, Yubao. (2021). Effect of the introduction of mono-functional monomer on the electro-optic properties of reverse-mode polymer stabilised cholesteric liquid crystal. *Liquid Crystals, 48*(8), 1162-1174.

CHAPTER 2

Abstract

Smart glass technology is increasingly prevalent in modern applications. One specific type of smart glass, known as polymer dispersed liquid crystal (PDLC), is created by incorporating epoxy monomers with liquid crystals. This PDLC can be enhanced by integrating silver nanoparticles (AgNPs) in varying concentrations, which leads to different size distributions on the nanoscale. To characterize these silver nanoparticles, UV-visible spectroscopy is used, while sodium borohydride serves as a reducing agent to fine-tune their size. The electro-optical properties of PDLC films are evaluated using a liquid crystal detector and an Abbe refractometer to measure refractive indices. PDLC thin film technology presents multiple advantages over conventional LCD materials, including simpler manufacturing processes, the elimination of polarizers, and the absence of viewing angle constraints. It does not require stringent control over the spacing between substrates, which facilitates the fabrication of large-area displays. The fabrication of these thin PDLC films employs a method known as polymerization-induced phase separation (PIPS), involving mixtures of curing monomers, liquid crystals, hardeners, and silver nanoparticles. Research has shown that the presence of silver nanoparticles, particularly those sized between 50-60 nm, significantly influences the structure of the polymer network. The cross-linking density of the polymer network increases with a higher content of curing monomers containing -NH2 groups. This composition enables the film to transition between transparent and opaque states, expanding the potential applications for PDLC technology and enhancing performance in various optical devices.

The epoxy-based PDLC film study was prepared by using Barium titanate (BaTiO3) nanoparticles. Polymer-induced phase separation (PIPS) method was used to demonstrate the substantial and synthetic properties of the film. It was noted that with the addition of doping Barium titanate (BaTiO3) into the film rise time increased, the threshold voltage decreased, the the saturation voltage decreased, and the constant ratio became better (Kumari & Sinha, 2020).

Prepared polymer-dispersed liquid crystal (PDLC) devices are mostly utilized in smart glass technology. The silver nanoparticles with a size of 15-20nm with a wetting agent were injected in PDLC; in which the injecting agent changed the physical and electro-optical properties. When voltage is given to PDLC it becomes opaque at off-state it becomes transparent at on-state. In this paper, the electro-optical performance of devices was studied by doping a surfactant and inserting silver nanoparticles into it (Yan et al., 2020).

Analyzed nematic liquid crystal 4-trans-pentyl-cyclohexyl cyan benzene (5PCH) and functionalized (hexane thiol capped) AgNPs were studied. In this paper, the effect of nematic liquid crystal 4-trans-pentyl-cyclohexyl cyan benzene (5PCH) doped with AgNPs enhanced the E-O properties in a twisted nematic configuration. The nanocomposite of 5PCH with AgNPs was synthesized by ultra-sonication with different concentrations examined. In this review, lower doping concentration AgNPs in 5PCH in the twisted nematic cells improved the E-O and switching properties for better displays (SJ, Gupta, Kumar, & Manjuladevi, 2020).

The silver nanoparticles on the kinetics of diffusion of Holographic polymer dispersed liquid crystal (H-PDLC) were studied. The non-local diffusion kinetics holographic dynamics model was applied to cause diffusion between liquid crystal, monomer, and silver nanoparticles were studied. The different parameters of the model showed that the diffraction grating for H-PDLC can be improved with different concentrations of silver nanoparticles (Liu et al., 2019).

2.1 Materials and Methods

Silver (I) nitrate (AgNO3) (99.99% pure) and sodium tetra hydroborate (NaBH4) (99.99% pure) were purchased from Tokyo Chemical Industry Co., Ltd.,) for AgNPs synthesis. Figure 1 shows the structures of the materials. In this study, the LC used NLC (NLC-1717, Anhui Hengyuan Chemical Co., Ltd.,) refractive index (RI) is n20/D 1.532. Di-functional chain length diglycidyl ethylene glycol (EDGE) resin RI is n20/D 1.463.

The heat-curable epoxy monomers are diglycidyl bisphenol A ether (BAGDE) and neopentyl glycol diglycidyl ether (NGDE) monomer RI is n20/D 1.457. The triethylenetetramine (TETA) hardener is n20/D 1.496 refractive indexes, and which were purchased from by Synasia (Suzhou) Co., LTD., The diethylenetriamine (DETA) hardener and Pentaerythritol-tetrakis-3-mercapto propionate (PERTMP) hardener, RI is n20/D 1.531

purchased from Shijiazhuang Yongsheng Huatsing Liquid Crystal Co., Ltd., were used. The chemical structures of these materials are shown in Figure 2-1,2-2 Without cleansing, we used the compounds.

The following chemicals were used in it:

EGDE/DGEBA/NGDE/TETA/DETA/PERTMP and NLCs for 7.0 hours at 70°C. The chemical structures of monomers, liquid crystals, and hardeners are given in Figure. 2-1,2-2.

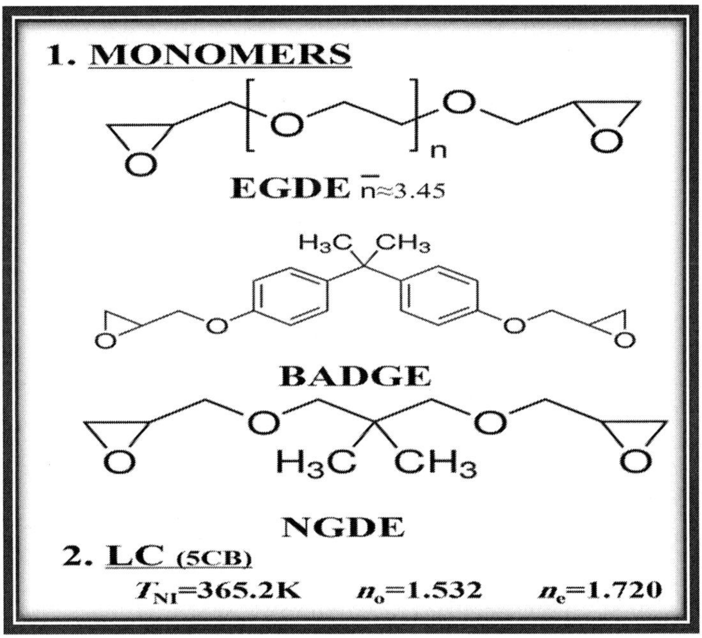

Figure 2-1 showing monomers and liquid crystal structures.

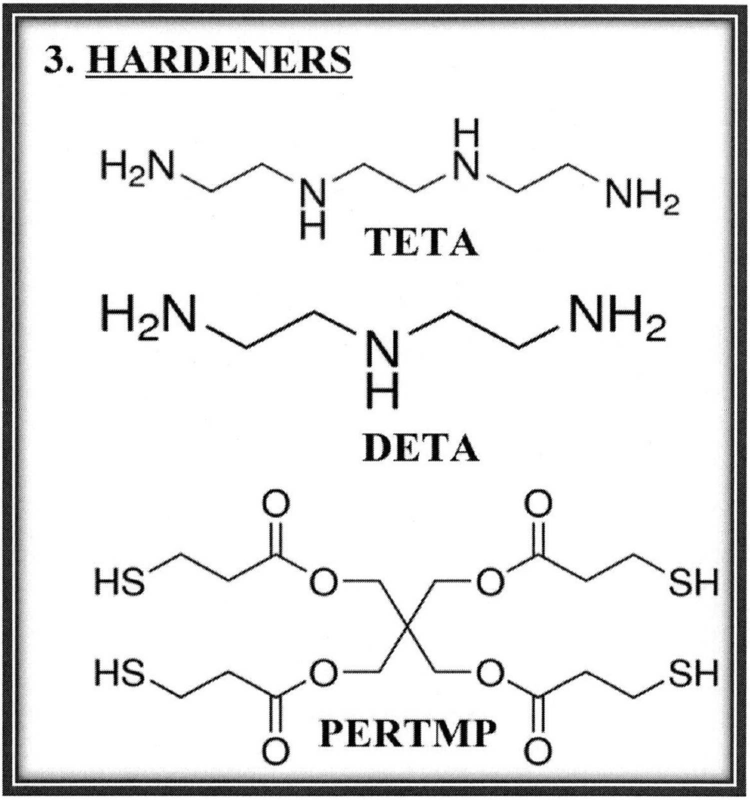

Figure 2-2 showing structures of Hardeners

2.1.1 Preparation method of the PDLC films

2.1.2 Synthesis of silver nanoparticles (AgNPs)

The test was performed in the dark to forestall the photodecomposition of AgNO3.Silver nitrate had initiated the usage of sodium tetra hydroborate (NaBH4) because of the lowering agent and capping agent, respectively specified mol ratios. Then AgNO3 dropwise changed into introduced right into a homogeneous combination of NaBH4 at consistent magnetic stirring, such that a molar ratio of 1:2 changed into maintained among AgNO3 and NaBH4.Instantaneously, the combination grew to become a mild yellow and lastly to a darkish wine crimson signifying the validity of AgNPs.

2.1.3 Preparation of PDLC films

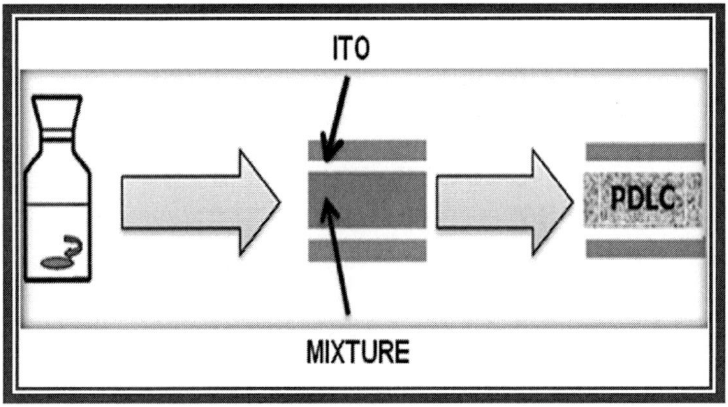

Figure 2-3 A schematic diagram of the preparation of PDLC films.

2.1.4 The epoxy curing process and Chemical reaction

Although there are numerous cure mechanisms, the one used most nowadays within the PDLC films business is the polyaddition chemical action with primary and secondary amines (M Ellahi, Gao, & Rafique, 2013). The key player within the hardening/natural/action/activity process is the strained oxirane ring, which opens up within the presence of amine-activated element, as shown in (Fig. 2-4, 2-5, and 2-6). Step Growth & Chain Growth Addition Polymerization Reaction.

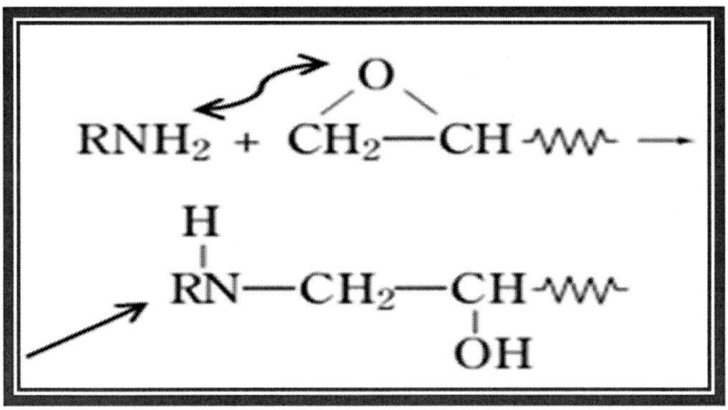

Figure 2-4 Primary amines undergo an addition reaction with the epoxide group to form a hydroxyl group and a secondary amine

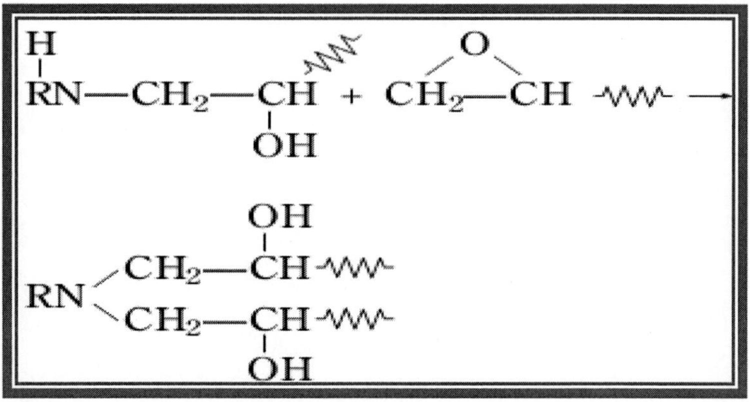

Figure 2-5 The secondary amine can further react with an epoxide to form a tertiary amine and an additional hydroxyl group

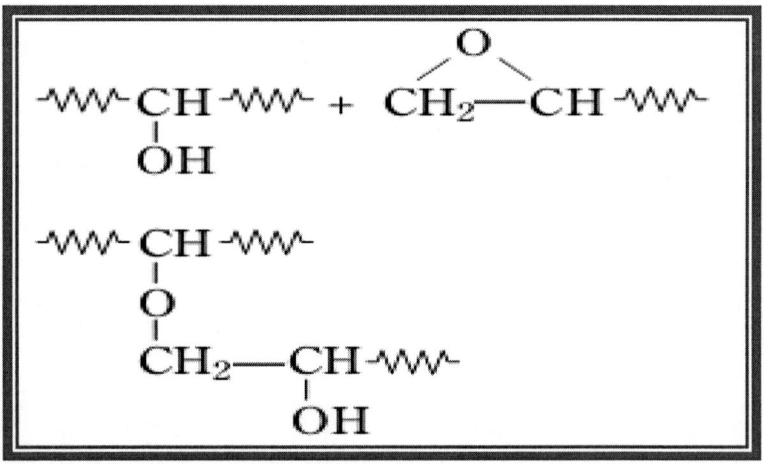

Figure 2-6 The presence of hydroxyls, however, has an important function because they assist in opening the epoxide ring.

The reaction happens while not the formation of by-products, and shrinkage is a smaller amount than five-hitter. Extremely reactive paraffin elements can allow the reaction to occur at below-close temperatures. Robust adhesive properties square measure because of the supply of gas and atoms within the compound, that covalent bond with the substrate. (Darla et al., 2014). The result is a highly cross-linked, three-dimensional supermolecule, schematized in (Fig. 2-7 and 2-8)

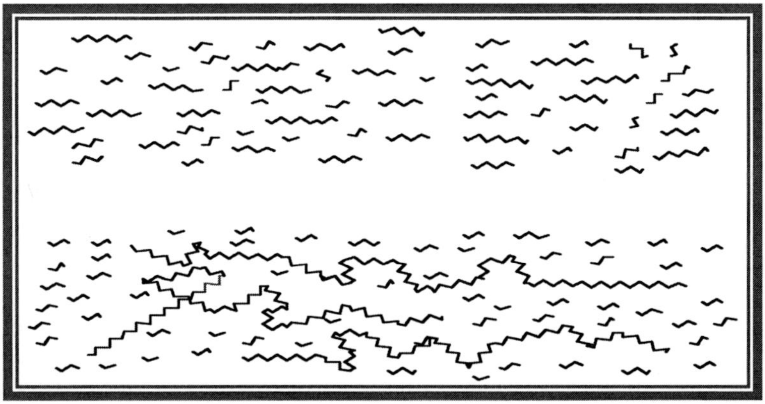

Figure 2-7 A schematic diagram of polymerization addition reaction of epoxy resin with a polyamine.

The linkages between adjacent compound chains turn out a rigid structure with glorious mechanical and thermal stability, a characteristic common to any or all thermosets as shown in (Fig.2-7 and 2-8).

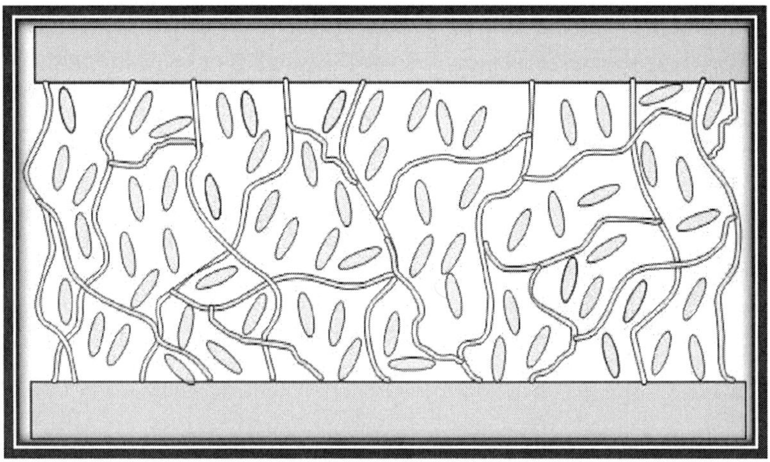

Figure 2-8 A schematic diagram of adjacent polymer chains of epoxy resin with a polyamine

PDLC doped with silver nanoparticles was obtained at the optimum concentration of silver nanoparticles. The characterization was done by surface plasma resonance by spectrophotometer, and field emission scanning electron microscopy (FE-SEM) showed that improvement in response time and reduction in switching voltage was achieved. The PDLC electro-optical properties doped with Ag-NPs, experimental results were based on the switching effect by changing ordinary amplitude modulation (AM) to frequency modulation (FM) when applied to the PDLC in electro-optical devices. The doping material changed the physical and chemical properties of liquid crystals. Frequency modulation (FM) is a new advancement in PDLC devices that was achieved for future usage of the technique (M. Zhang et al., 2013).

This paper revealed that the AgNPs in PDLC induced by the phase separation method showed two resonance peaks at the interface; on the other hand, by increasing the incident angle, the resonance effect showed a long wavelength with a blue shift by changing the position of NPs there was no change in resonance peaks. The first resonance effect, localized-surface plasmon resonance, impacts the electro-optical properties of H-PDLC grating (Gui et al., 2015).

The author, anticipating the study of methacrylate's monomer, provided improved results of super electro-optical properties compared to other acrylate monomers. The methacrylate monomer cross-linking and its rigidity provided better display devices. The experimental work was done on different acrylate monomers and variable weights of liquid crystals. The characterization results showed improved morphology and physical properties of PDLC film (Chen et al., 2006).

The author analyzed the oxygen-containing heterocyclic monomer induced into the polymer matrix. The effect of oxygen-bearing heterocyclic monomers, cross-linker chain length-monomer, and cell thickness relatively improves the morphology, high contrast ratio, less response time, and less switching voltage. This work provides an application of PDLC with high contrast ratios at the slimmer film (Saeed et al., 2021).

The PDLC film was made up of acrylate monomers, liquid crystals, and Indium tin oxide (ITO) nanoparticles with different weights. The ITO NPs were prepared with 3-methacrylate oxypropyl trimethoxysilane (KH570) by ultra-sonication of the mixture. The NIR absorption property of PDLC film has been increased the electro-optical properties were reduced. In the future, we can adjust the ITO NPs and obtain PDLC film with good electro-optical properties (C. Zhang et al., 2020).

The work utilized a model based on a coupled partial differential equation to analyze large liquid crystal devices with non-uniform optical and electrical signal distributions. The simulation compared the frequencies of optical and electrical signals under applied voltage, revealing that sinusoidal waves effectively operate display devices at higher frequencies (Gao et al., 2021).

The polymer-stabilized liquid crystal (PSCL) system incorporated a bi-functional monomer, offering unique properties compared to the traditional mono-functional monomer used in reverse phase-stabilized cholesteric liquid crystals (RPSCLC). Introducing a non-functional monomer into the bi-functional monomer RPSCLC led to notable enhancements. Specifically, a higher concentration of the mono-functional monomer resulted in an increased contrast ratio, reduced hysteresis, and lower driving voltage. This represents a significant advancement in the research of PSLC, paving the way for improved performance in liquid crystal devices (Zhao et al., 2021).

The ionic stabilizer and crosslinking agent were applied to a glass surface that had silver nanoparticles (AgNPs) deposited on it. The ionic silsesquioxane served as a stabilizer, forming a network on the glass, silica, and alumina surfaces. Staphylococcus aureus formed colonies on the glass plate, even after being sterilized three times. As a result, the silsesquioxane-coated glass plate can be developed for use as an antibacterial film (Schneid et al., 2015).

In this article, the author explored the application of nanoparticles (NPs) on rough surfaces, which is a departure from the common focus on smooth surfaces. The study included various statistical parameters related to the deposition of NPs under non-ideal conditions. The nanoparticles were characterized using atomic force microscopy (AFM), revealing that data processing algorithms play a crucial role in addressing uncertainties and systematic errors associated with AFM measurements of NPs. This highlights the importance of careful analysis in understanding the behaviour of nanoparticles in diverse environments (Klapetek, Valtr, Nečas, Salyk, & Dzik, 2011).

A tri-block copolymer amphiphilic (MAM) was introduced into epoxy resin, leading to an improvement in the hardness of the material. While this amphiphilic copolymer does not alter the curing process, it appears to hinder it slightly. The study examined the mechanical properties, including tensile strength, thermal behaviour, and phase structure. Notably, the addition of MAM increased the hardness of the epoxy resin without decreasing the glass transition temperature, and the structural integrity of the epoxy remained unaffected by the incorporation of the copolymer. This suggests that MAM can enhance the performance of epoxy resins while maintaining their fundamental properties (Tao et al., 2020).

Silver nanoparticles were synthesized using the inert gas aggregation method, which allows for the controlled aggregation of clusters (such as trimers and dimers) into non-cluster sources. This approach capitalizes on a well-documented physical phenomenon in nanoparticle preparation. High-resolution transmission electron microscopy revealed that the resulting nanoparticles exhibited an icosahedral shape, oriented with a fivefold symmetry axis angle perpendicular to the substrate. By effectively controlling the size and shape of the metal nanoparticles, the inert gas aggregation method proves to be an excellent technique for the preparation of silver NPs (Gracia-Pinilla, Ferrer, Mejía-Rosales, & Pérez-Tijerina, 2009).

The investigation focused on silver nanoparticles (AgNPs) functionalized with local anaesthetics such as procaine, dibucaine, or tetracaine. A coating layer was formed through the adsorption of asparagine and silica onto the nanoparticles. This functionalization enhances the biological application potential of AgNPs, as the combination of silver with local anaesthetics provides a material with notable bactericidal properties. This approach could lead to advancements in the development of antibacterial therapies, leveraging the unique properties of both the silver nanoparticles and the local anaesthetics (Mohan, 2013). The paper reviewed various methods for modifying epoxy resin, highlighting that the modified epoxy demonstrates excellent thermal properties, adhesion, electrical conductivity, and toughness. One significant feature is its ability to transition from a liquid to a solid state, enhancing its utility. These properties make modified epoxy resin highly applicable in the engineering field, where it can be used for diverse engineering purposes. Different methods have been explored for the modification process, contributing to its improved performance and versatility (Mocanu et al., 2013).

2.2 Preparation of PDLC without silver Nanoparticles

All chemicals were used as received. The chemical compositions of epoxy monomers / LC mixtures/hardeners are given in Table 2-1, 2-2, 2-3 and 2-4 for 7.0 h at 70 °C temperature.

Table 2-1 shows the composition of sample A1-A4

Samples	Epoxy Monomers Mixture（60 wt%） 7.0 Hours at 348.15K		NLC-1717/ wt%
	TETA wt%	EGDE / BADGE wt%	
A1	05	30/25	40
A2	05	29/26	40
A3	05	28/27	40
A4	05	27/28	40

Table 2-2 shows the composition of sample B1-B4

Samples	Epoxy Monomers Mixture（60 wt%） 7.0 Hours at 348.15K		NLC-1717/ wt%
	DETA wt%	EGDE / BADGE wt%	
B1	05	30/25	40
B2	05	29/26	40
B3	05	28/27	40
B4	05	27/28	40

Table 2-3 shows the composition of sample C1-C4

Samples	Epoxy & Thiol Monomers Mixture (60 wt%) 7.0 Hours at 348.15K		NLC-1717/ wt%
	PERTMP wt%	EGDE / BADGE wt%	
C1	05	30/25	40
C2	05	29/26	40
C3	05	28/27	40
C4	05	27/28	40

Table 2-4 shows the composition of sample D1-D4

Samples	Epoxy & Thiol Monomers Mixture (60 wt%) 7.0 Hours at 348.15K		NLC-1717/ wt%
	PERTMP wt%	EGDE / NGDE wt%	
D1	05	25/30	40
D2	05	26/29	40
D3	05	27/28	40
D4	05	26/29	40

2.3 Preparation of PDLC with silver Nanoparticles (AgNPs)

Table 2-5 shows the composition of sample E1-E4

Samples	Epoxy Monomers Mixture (60 wt%) 7.0 Hours at 348.15K		NLC-1717/ wt%
	TETA wt%	EGDE / BADGE/AgNPs wt%	
E1	05	29.75/25/0.25	40
E2	05	29.50/25/0.50	40
E3	05	29.25/25/0.75	40
E4	05	29.00/25/1.00	40

Table 2-6 shows the composition of sample E1-E4

Samples	Epoxy Monomers Mixture (60 wt%) 7.0 Hours at 348.15K		NLC-1717/ wt%
	DETA wt%	EGDE / BADGE/AgNPs wt%	
F1	05	29.75/25/0.25	40
F2	05	29.50/25/0.50	40
F3	05	29.25/25/0.75	40
E4	05	29.00/25/1.00	40

Table 2-7 shows the composition of sample G1-G4

Samples	Epoxy & Thiols Monomers Mixture (60 wt%) 7.0 Hours at 348.15K		NLC-1717/ wt%
	PERTMP wt%	EGDE / BADGE/AgNPs wt%	
G1	05	29.75/25/0.25	40
G2	05	29.50/25/0.50	40
G3	05	29.25/25/0.75	40
G4	05	29.00/25/1.00	40

Table 2-8 shows the composition of sample H1-H4

Samples	Epoxy & Thiols Monomers Mixture (60 wt%) 7.0 Hours at 348.15K		NLC-1717/ wt%
	PERTMP wt%	EGDE / NGDE/AgNPs wt%	
H1	05	29.75/25/0.25	40
H2	05	29.50/25/0.50	40
H3	05	29.25/25/0.75	40
H4	05	29.00/25/1.00	40

The study examined the morphology and optical properties of liquid crystal composites using the polymerization-induced phase separation (PIPS) method. Two types of composites were analyzed: polymer dispersed liquid crystals (PDLC) and phase-separated composite films (PSCOF), prepared with specific mass ratios. It was found that the exposure intensity directly impacts the electro-optical properties and morphology of the PDLCs. In contrast, the exposure temperature was identified as a critical factor influencing the properties of the PSCOFs. This nuanced understanding of the effects of exposure conditions helps in tailoring the performance characteristics of these materials for various applications (K. Li et al., 2020).

The investigation focused on incorporating silver-coated polystyrene microspheres (Ag-coated PSMS) into polymer-dispersed liquid crystals (PDLC). The silver was applied to the polystyrene using a chemical method, resulting in the induction of an electric field when an external electric field was applied. This induced electric field notably lowered the driving voltage required for smart display applications. Additionally, experimental work was conducted with varying weight percentages of the pre-polymer and liquid crystal to improve the optical characteristics of the PDLC materials. These modifications aim to enhance the efficiency and adaptability of PDLCs in practical applications (Chan et al., 2016).

The PDLC was prepared by synthesizing nematic liquid crystals with polyester resin in the laboratory. Experimental results demonstrated favourable electro-optical properties, highlighting the potential of the material in smart display applications. Further efforts were made to enhance the inducing materials within the PDLC system to optimize performance. This ongoing development aims to improve the effectiveness and versatility of PDLCs in various practical uses (Kashima et al., 2010). The photochemical polymerization method was employed for display. This approach involved varying the weight percentages of the cross-linker, ultraviolet-curable monomer, initiator, and liquid crystal, allowing for modifications to the chain length of the cross-linker. Utilizing lower concentrations of liquid crystal resulted in enhanced electro-optical properties, ultimately leading to a higher quality material for display applications. This careful adjustment of components highlights the intricate balance required to optimize the performance of the display technology (Mormile et al., 2000).

References

Abbasifar, Ahmadreza, Ghani, Sepideh, Ahsani Irvani, Maryam, Rafiee, Behnam, Valizade Kaji, Babak, & Akbari, Alireza. (2017). Antibacterial activity of silver nanoparticles synthesized by using extracts of Hedera helix. *Zahedan Journal of Research in Medical Sciences, 19*(1),5920-5923.

Ahmad, Farzana, Jamil, 1 M, & Jeon, YJ. (2014). Current trends in studies on reverse-mode polymer dispersed liquid-crystal films—A review. *Electronic Materials Letters, 10*(4), 679-692.

Ahmad, Farzana, Jamil, Muhammad, Jeon, Young Jae, Woo, Lee Jin, Jung, Jae Eun, & Jang, Jae Eun. (2012). Investigation of nonionic diazo dye-doped polymer dispersed liquid crystal film. *Bulletin of Materials Science, 35*(2), 221-231.

Ahmad, Farzana, Jamil, Muhammad, Jeon, Young Jae, Woo, Lee Jin, Jung, Jae Eun, Jang, Jae Eun, Park, Jinsang. (2011). Comparative study on the electrooptical properties of polymer-dispersed liquid crystal films with different mixtures of monomers and liquid crystals. *Journal of Applied Polymer Science, 121*(3), 1424-1430.

Bouteiller, Laurent, & Barny, P LE. (1996). Polymer-dispersed liquid crystals: preparation, operation and application. *Liquid crystals, 21*(2), 157-174.

Chan, Chia-Hua, Wu, Ting-Yun, Yen, Meng-Hua, Lin, Chu-En, Cheng, Ko-Ting, & Chen, Chii-Chang. (2016). Low power consumption and high-contrast light scattering based on polymer-dispersed liquid crystals doped with silver-coated polystyrene microspheres. *Optics express, 24*(4), 29963-29971.

Chandrasekhar, S. (1978). X–ray analysis of liquid crystals. *Current Science, 47*(15), 523-525.

Capricho, Jaworski C, Fox, Bronwyn, & Hameed, Nishar. (2020). Multifunctionality in epoxy resins. *Polymer Reviews, 60*(1), 1-41.

Castellón, Erick, & Levy, David. (2018). Smart Windows Based on Liquid Crystal Dispersions. *Transparent Conductive Materials: Materials, Synthesis, Characterization, Applications,Weinheim,Germany,5*(4), 37-365.

Chen, Tien-Jung, Chen, Yu-Fan, Sun, Chia-Hsing, & Wu, Jin-Jei. (2006). Electro-optical properties of reverse-mode films of planar aligned polymer-dispersed liquid crystal. *Journal of Polymer Research, 13*(2), 85-89.

Chung, Seok-Hwan, Gil, Eun-Kyung, Noh, Hee-Yeon, Jeong, Soon Moon, & Choi, Byeongdae. (2014). Effect of Urethane Acrylate Oligomer and UV Curing Temperature on Electro-Optical Property of Polymer-Dispersed Liquid Crystal. *Molecular Crystals and Liquid Crystals, 601*(1), 1-10.

Darla, Mallikharjuna Rao, Hegde, Sriharsha, & Varghese, Soney. (2014). Effect of BaTiO 3 nanoparticle on electro-optical properties of polymer dispersed liquid crystal displays. *Journal of Crystallization Process and Technology,4*(1), 60-63.

D Jákli, A. (2019). One-and two-dimensional fluids: Properties of smectic, lamellar and columnar liquid crystals: (1st eds.) England, UK.CRC Press, Jákli, A. 352.

Doane, JW, Vaz, NA, Wu, B-G, & Žumer, Slobodan. (1986). Field controlled light scattering from nematic microdroplets. *Applied Physics Letters, 48*(4), 269-271.

Ellahi, M, & Rafique, MY. (2016). Electro-optical properties of PDLC films using diethylenetriamine (DETA) hardener. *Molecular Crystals and Liquid Crystals, 638*(1), 103-110.

Fergason, James L. (1985). Polymer encapsulated nematic liquid crystals for display and light control applications. *SID Dig. Tech. Paper, 16*, 68-70.

Huang, Guang-Chun, & Lee, Jong-Keun. (2011). Cure Kinetics and Dynamic Mechanical Properties of an Epoxy/Polyoxypropylene Diamine System. *Polymer Korea, 35*(3), 196-202.

Hui, LI, Feifan, YANG, Jing, PENG, Fan, PAN, Yuntao, WU, Zhang, Yanduo, & Xiaolin, XIE. (2021). Electro-Optical Properties of a Transmissive Diffuser based on Electrically-Tuneable Polymer-Dispersed Liquid Crystal Film. *Materials Science, 27*(1), 8-12.

Kashima, Miki, Cao, Hui, Liu, Huijing, Meng, Qingyong, Wang, Ding, Li, Fasheng, & Yang, Huai. (2010). Effects of the chain length of crosslinking agents on the electro-optical properties of polymer-dispersed liquid crystal films. *Liquid Crystals, 37*(3), 339-343.

Kato, Shinji, Chen, Feng-Qi, & Pac, Chyongjin. (2004). Anchoring effects of self-assembled monolayers for polymer-dispersed liquid crystal films. *The Journal of Physical Chemistry B, 108*(1), 320-326.

Khan, Raj Kumar, & Ramarao, Pratibha. (2021). Selective stabilization of blue phase liquid crystals using spherical and rod-shaped colloidal nanocrystals. *Journal of Applied Physics, 129*(2), 1-15.

Mormile, P, Musto, P, Petti, L, Ragosta, G, & Villano, P. (2000). Electro-optical properties of a PDLC based on unsaturated polyester resin. *Applied Physics B, 70*(2), 249-252.

Önsal, Gülnur, Kocakülah, Gülsüm, Kahyaoğlu, Aliye, & Köysal, Oğuz. (2019). Influence of azo dye concentration on dielectric response in polymer dispersed liquid crystal composites. *Journal of Molecular Liquids, 284*(15), 607-615.

Paluvai, Nagarjuna Reddy, Mohanty, Smita, & Nayak, SK. (2014). Synthesis and modifications of epoxy resins and their composites: a review. *Polymer-Plastics Technology and Engineering, 53*(16), 1723-1758.

Zhang, Menghua, Zheng, Jihong, Gui, Kun, Wang, Kangni, Guo, Caihong, Wei, Xiaopeng, & Zhuang, Songlin. (2013). Electro-optical characteristics of holographic polymer dispersed liquid crystal gratings doped with nanosilver. *Applied Optics, 52*(31), 7411-7418.

Zhao, Rui, Li, Xiaoshuai, Wang, Kemeng, Huai, He, Ma, Hongmei, & Sun, Yubao. (2021). Effect of the introduction of mono-functional monomer on the electro-optic properties of reverse-mode polymer stabilised cholesteric liquid crystal. *Liquid Crystals, 48*(8), 1162-1174.

CHAPTER 3

Abstract

Smart glass technology is increasingly prevalent in modern applications. One specific type of smart glass, known as polymer dispersed liquid crystal (PDLC), is created by incorporating epoxy monomers with liquid crystals. The PDLC can be enhanced by integrating silver nanoparticles (AgNPs) in varying concentrations, which leads to different size distributions on the nanoscale. To characterize these silver nanoparticles, UV-visible spectroscopy is used, while sodium borohydride serves as a reducing agent to fine-tune their size. The electro-optical properties of PDLC films are evaluated using a liquid crystal detector and an Abbe refractometer to measure refractive indices. PDLC thin film technology presents multiple advantages over conventional LCD materials, including simpler manufacturing processes, the elimination of polarizers, and the absence of viewing angle constraints. It does not require stringent control over the spacing between substrates, which facilitates the fabrication of large-area displays. The fabrication of these thin PDLC films employs a method known as polymerization-induced phase separation (PIPS), involving mixtures of curing monomers, liquid crystals, hardeners, and silver nanoparticles. Research has shown that the presence of silver nanoparticles, particularly those sized between 50-60 nm, significantly influences the structure of the polymer network. The crosslinking density of the polymer network increases with a higher content of curing monomers containing -NH2 groups. This composition enables the film to transition between transparent and opaque states, expanding the potential applications for PDLC technology and enhancing performance in various optical devices.

Analysis of multifunctional epoxy resin was studied by the multifunctional epoxy which has the property to become useful in the ionic field, electrically conductive resins, high-temperature resins, recyclable resin, rapid cure resins, flexible and formable resin, and toughness enhanced resins. These all are new advancements in multifunctional epoxy resin. (Chen et al., 2006). PDLC film was obtained in reverse mode using a mixture of three different kinds of monomers and nematic liquid crystals. The alignment of the mixture is achieved by rubbing in the direction of the sample during the glass fabrication process. When in the on-state, the film becomes opaque as the refractive index of the liquid crystal matches that of the polymer. In the off-state, however, there is a mismatch, causing the film to become transparent. It's important to control the curing temperature and time for this type of PDLC to ensure optimal performance (Capricho et al., 2020).

The photopolymerization process of PDLC involved adding reactive 2-Ethyl hexyl acrylate monomer alongside an aliphatic urethane acrylate (UA) oligomer to the pre-polymer mixture. The concentration of the urethane acrylate (UA) plays a crucial role, affecting factors such as driving voltage, response time, contrast ratio, and droplet size within the film. Additionally, both the UV curing temperature and the concentration of the UA oligomer have a direct impact on the optical properties of the PDLC film. This interplay of variables is essential for optimizing the film's performance and characteristics (Chung, Gil, Noh, Jeong, & Choi, 2014).

The multilayered structure of Zito/Ag/ZITO, consisting of layers with sizes of 100 nm, 8 nm, and 42 nm, was integrated into an acrylate-based dispersed liquid crystal polymer. This transparent multilayered film exhibits notable transmittance and sheet resistance when exposed to UV light. The electro-optical properties, along with the morphology and photopolymerization kinetics of the film, were thoroughly analyzed using conducting electrodes. As a result, the PDLC film formed contains liquid crystal droplets with an average microstructure size ranging from 2 to 5 μm (Huang & Lee, 2011).

Epoxy monomers have found widespread applications across various industries, including aerospace, automotive, agriculture, and engineering. A key benefit of using renewable source-based epoxy is its positive impact on the environment, helping to mitigate global warming. Traditionally, bisphenol-A, which is recognized as a toxic chemical, has been replaced by di phenolic acid in these formulations. Additionally, many curing agents have transitioned to bio-based chemicals and inorganic catalysts, enhancing the sustainability of these materials. The bio-based epoxies not only contribute to ecological benefits but also offer high-performance implementations in their respective applications (Paluvai et al., 2014). The development of PDLC (Polymer-Dispersed Liquid Crystal) film with reduced driving voltage can be enhanced by adjusting the concentration of acrylate monomer, specifically A3DA. This monomer features a molecular structure that includes a benzene ring, a dodecyl tail, and a propyl spacer, which contributes to the material's overall properties. Notably, the incorporation of this modified composition has led to significant improvements in the electro-optical characteristics of the PDLC film, achieving a notable decrease in driving voltage of approximately 0.55 V/μm. Characterization of the new PDLC film was performed using confocal laser scanning microscopy, which provided insights into its structure and performance. This advancement in PDLC materials highlights the potential for creating more efficient and effective electro-optical devices (Kizhakidathazhath et al., 2020).

The UV-cured PDLC was synthesized using the polymer-induced phase separation (PIPS) method. In this process, certain portions of the liquid solidified while some pre-polymers remained uncured within the liquid crystal droplets. It was noted that diffusion across the boundaries was not fully completed. As a result, the electro-optical properties of the UV-cured PDLC exhibit changes over time, leading to a transient phenomenon that is noticeable within two weeks. This time-dependent behaviour highlights the dynamic nature of the material and its properties (Hakemi et al., 2020). The reverse-mode PDLC film consists of patterned polymer walls encasing the liquid crystal. When an electric voltage is applied, the liquid crystal molecules reorient, causing the film to turn milky. Conversely, in the off-state, the film becomes transparent. The optical characteristics of the film improve significantly when the optimal nematic liquid crystal is selected. This selection plays a crucial role in enhancing the performance and clarity of the film in different states (Huihui Wang et al., 2015).

3.1 UV–Vis spectral analysis

In metals, specifically with silver nanoparticles, the conduction band and valence band are closely aligned, allowing for unoccupied electrons to revolve freely. When these electrons move upward towards the Surface Plasmon Resonance (SPR) band, they are influenced by the resonant oscillation created by the silver nanoparticles. The electric field wave induces a polarization of the electrons, as the ionic centres of the silver are surrounded by the nanoparticles. This configuration results in a biphasic oscillation of electrons, leading to increased absorption properties.

The absorption of light in metal nanoparticles, particularly silver, directly influences their observed colour. This absorption is indicative of the chemical environment surrounding the nanoparticles. The size of the particles is predominantly determined by sodium tetra hydroborate, which acts as a reducing agent and strongly influences particle size formation. When light is absorbed in frequencies that resonate with the Localized

Surface Plasmon Resonance (LSPR), it alters the colour of the metal due to this specific light absorption. Additionally, as the LSPR gets excited, light scattering occurs, leading to notable changes in the colour of light, which can give the nanoparticles a luminous, almost fluorescent appearance when they are dispersed in a solution (Masutani et al., 2007).

UV-visible spectroscopy serves as a crucial characterization technique for understanding the shaping and effects of nanoparticles. The shape of the spectrum, along with factors such as density, size, surrounding medium, free-electrons, and interparticle interactions, significantly influence the surface resonance of these nanoparticles. Monitoring the aggregation of nanoparticles (NPs) and electron injection is essential for this analysis. The absorption or extinction peaks observed between 390-410 nm exhibit varying intensity and bandwidth due to different lengths and length distributions of nanoparticles. According to quantum length theories, the resonance band becomes broader as the particle length decreases, highlighting the relationship between size and optical properties in these materials.

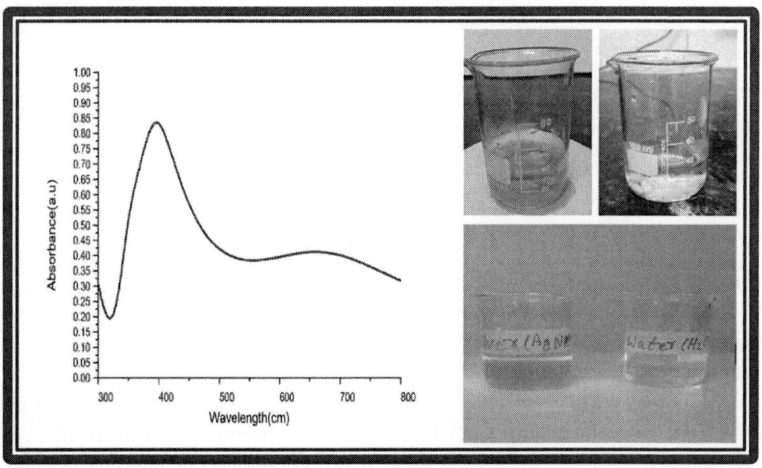

Figure 3-1 shows the UV-vis spectrum and color of AgNPs

UV–Vis spectroscopic analysis is the most important method to confirm the synthesis of nanoparticles. The absorbance spectrum of the sample was in the range of 300–800 nm, by using a RIGOL Ultra- 3660. In the UV-Vis spectrometer, distilled water was used as a reference. Figure 3-1 shows the AgNPs solution by formation of the nanoparticles.

The UV–visible absorption spectra shown in the graph of the AgNPs peak at ≈ ((390-410 nm) were recorded. By a change in the colour of the reaction solution from colourless to pale yellow which can be visually observed formation of the AgNPs during the reduction process is indicated. In this picture, a laser light beam is passed through water and AgNPs immediately after nanoparticle solution making, Silver nanoparticles in solution give a colloid. The light is scattered when shining on the colloid, showing the path of light. The laser passes from water but scatters at the AgNP solution which is confirmation of nanoparticle formation.

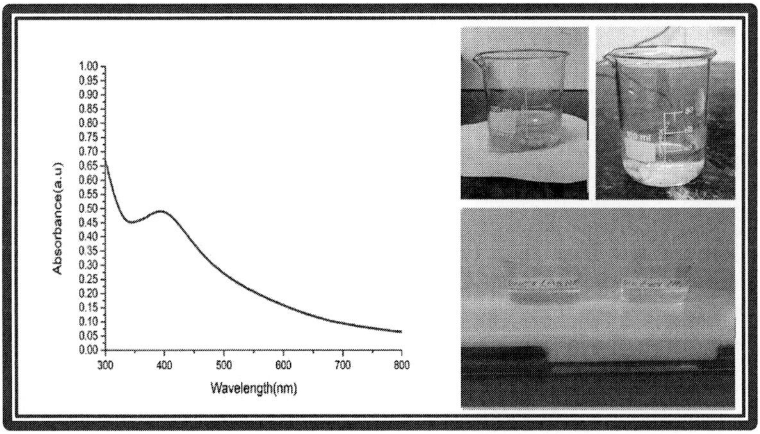

Figure 3-2 shows the UV-vis spectrum and color of AgNPs

Figure 3-2 shows the AgNP solution with and without the formation of the nanoparticles. The AgNP peak is at ≈ (390-415nm) as shown in the graph. In this picture, a laser light beam is passed through water and AgNPs. The laser light passes from water but scatters at the AgNP solution. This is confirmation of nanoparticle formation.

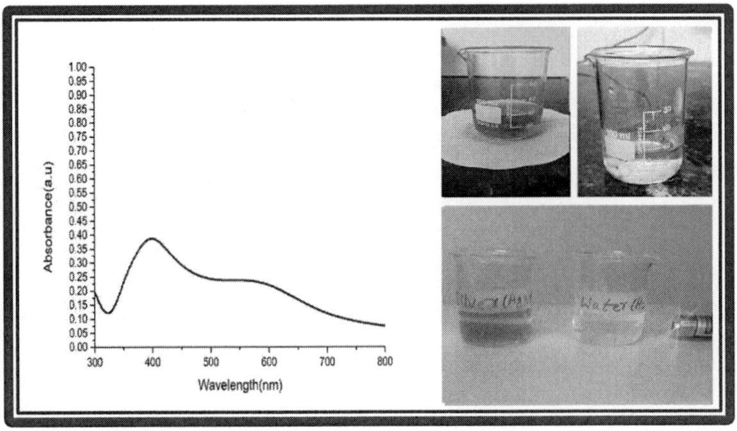

Figure 3-3 shows the UV-vis spectrum and color of AgNPs

Figure 3-3 shows the formation of the nanoparticle. The UV–visible absorption spectra showed the AgNPs ≈ (395-420 nm) were shown in the graph. In this picture, a laser light beam is passed through water and AgNps. The laser passes from water but scatters at the AgNP solution. This is confirmation of nanoparticle formation.

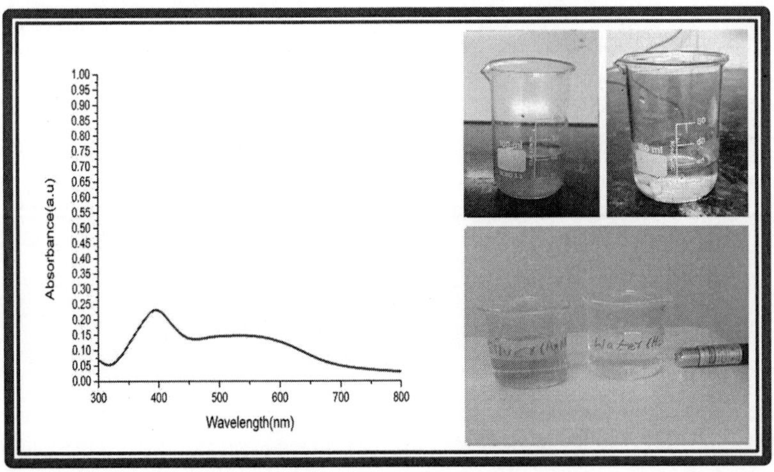

Figure 3-4 shows the UV-vis spectrum and color of AgNPs

Figure 3-4 shows the formation of the nanoparticle. The UV–visible absorption spectra showed the AgNPs ≈ (400-425 nm) as shown in the graph. In this picture, a laser light beam is passed through water and AgNPs. The laser passes from water but scatters at the AgNP solution. This is confirmation of nanoparticle formation.

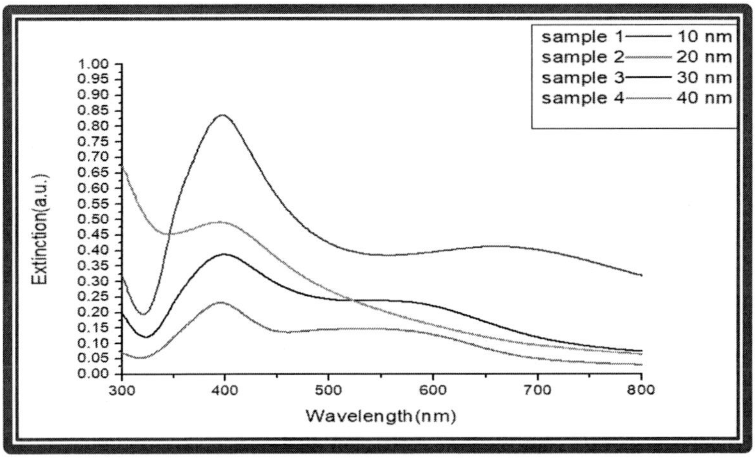

Figure 3-5 UV-Vis extinction spectra and the size of AgNPs

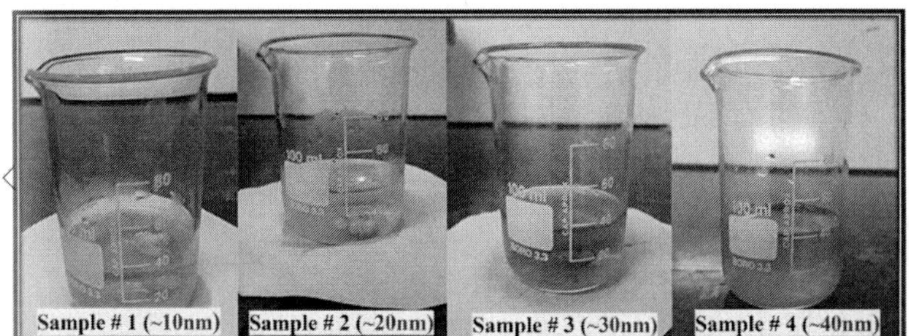

Figure 3-6 the distinctive color of different sized silver nanoparticles.

The figure shows the UV-visible spectroscopy measurement of plasmon resonance spectra of different sizes of silver nanoparticles. The size of silver nanoparticles is related to sodium tetra hydroborate concentration which gives particle size determination. Nanoparticle size usually starts from 1-

100nm as experimental data shows the absorbance peaks increase from 396-403nm and broaden in width.

3.2 Electro-Optical measurement of PDLC films

The epoxy resin can be modified through different curing processes, which largely depend on the types of curing agents used. By incorporating rubbery liquid materials into the resin, the toughness, mechanical strength, and physical and thermal properties of the epoxy can be significantly improved. These modifications lead to the development of tougher epoxy formulations that demonstrate enhanced performance in various applications (Unnikrishnan & Thachil, 2006).

The chemical reduction method was employed to prepare nanoparticles. To prevent the aggregation of particles, polyvinyl pyrrolidone (PVP) was added to the material containing sodium borohydride and silver. Various samples were prepared with different concentrations of silver. To characterize the silver nanoparticles, techniques such as UV-Vis spectroscopy, transmission electron microscopy (TEM), and dynamic light scattering (DLS) were utilized (Mehr, Khanjani, & Vatani, 2015).

The diffusion method was utilized to react silver nanoparticles (AgNPs) with the plant extract of Hydra helix. The resulting AgNPs demonstrated microbicidal activity against Klebsiella pneumonia and Bacillus subtilis. The characterization of the AgNPs was conducted using transmission electron microscopy (TEM), UV-Vis spectroscopy, and dynamic light scattering (DLS). It was observed that both silver nitrate and the raw plant extract exhibited no antibacterial activity, which led to the decision to employ a biological method for the preparation of the nanoparticles (Abbasifar et al., 2017).

The studies focused on doping zinc oxide nanoparticles (ZnO NPs) into polymer-dispersed liquid crystals (PDLC). The results indicated that the morphology and electro-optical properties were significantly influenced by the presence of ZnO NPs. Specifically, doping with ZnO NPs resulted in an increased response time, as well as a decrease in both driving and threshold voltage. Additionally, the morphological analysis revealed that increasing the amount of doping led to an increase in the droplet size within the PDLC films (P. Zhang, Li, Yang, Feng, & Xie, 2012).

This paper discusses the fabrication of zinc acrylate nanoparticles, specifically detailing how zinc oxide (ZnO) nanoparticles were uniformly layered on an indium tin oxide (ITO) coated glass substrate. The fabrication technique used had a considerable influence on several electro-optical properties, such as driving voltage, contrast ratio, threshold voltage, and response time. Additionally, this method offers significant benefits over traditional fabrication processes due to its simplicity and time efficiency (Ahmad, Jeon, Jeon, & Jamil, 2021). The author studied the anchoring energy of the liquid crystal and its deformation by applying strain to it. When strain is applied anchoring energy decreases at the center and increases at the polymeric walls. The solvent-induced phase separation (SIPS) method was used for PDLC film preparation. This paper was based on a simulation of free anchoring energy in PDLC prepared by thermos tropic liquid crystal and polymethyl methacrylate coating over the film (Miller & Mocanu, 2000).

A Liquid Crystal device parameters tester (LCT-5016C, Changchun Liancheng Instrument Co. Ltd) has been used. The electro-optical (E-O) properties were measured using a liquid crystal detector to test the photoelectric properties of the sample. The sample was induced in a He-Ne laser at room temperature and in switching voltages in PDLC films (Schneid et al., 2015). The PDLC has very devastating Electro-optical properties. Optical properties highly depend on the epoxy, monomer, and liquid crystal-chemical nature. Additionally, the effect of epoxy chemical compound structure on PDLC films is understood by studying the electro-optical properties of sample A1-H4.

The liquid crystal droplet size determines the transmittance of PDLC film. The material surface transmittance is its capability to transmit radiant energy which is the ratio of electromagnetic power incident on a sample and pass through it. A distinction to the transmission constant is the magnitude relation of the transmitted to the incident field of force. Hence by regulating the monomer ratio the liquid crystal droplet size can easily be determined and transmittance increases with increasing the liquid crystal droplet size.

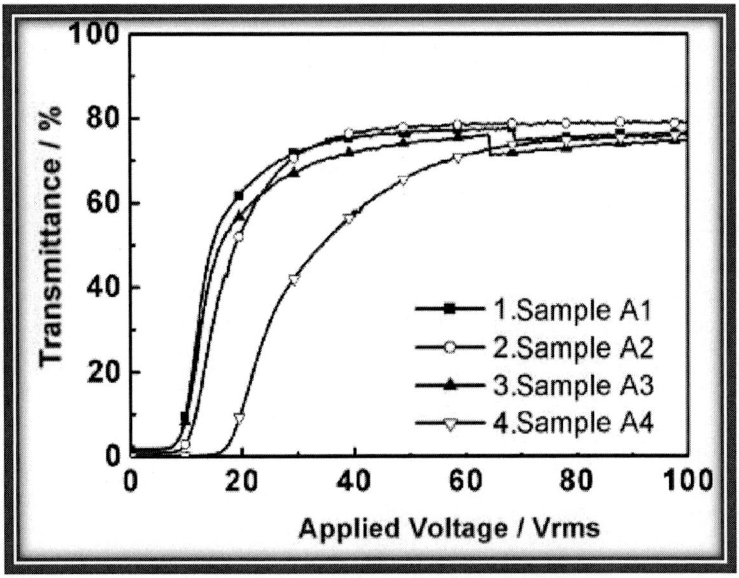

Figure 3-7 shows transmittance vs applied voltage of sample A1-A4.

The sample A1-A4 in (Figure 3-7) the transmittance of PDLC film as a function of an applied voltage. Expansion of applied voltage and the transmittance of the sample occur at 75%-80%. The sample A1-A4 shows the smooth saturation at 100 volts. While A1-A3 shows the moving smoothly then sharp downward movement occurs at 65-68 volts. This occurs due to the monomer concentration. As a result, the transmittance is achieved when the applied voltage is increased while the transmittance of the sample is obtained step by step.

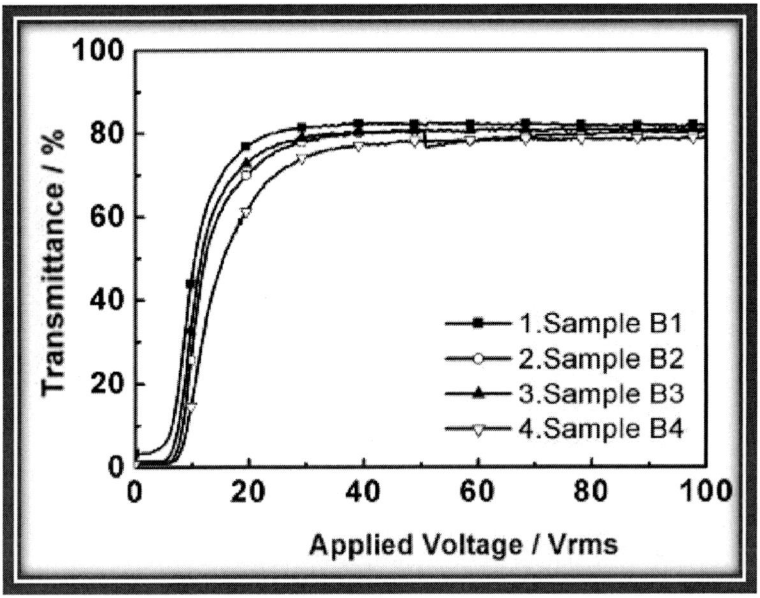

Figure 3-8 shows transmittance vs applied voltage of sample B1-B4.

(Figure 3-8) shows sample B1-B4 the transmittance of PDLC film as a function of an applied voltage. Expansion of applied voltage and the transmittance of the sample occur at 75%-85%. Sample B1-B4 shows the smooth saturation at 100 volts while B2 shows the peak moving smoothly then sharp downward movement occurs at approximately 52 volts. This occurs due to the monomer concentration. As a result, the transmittance is achieved when the applied voltage increases while the transmittance of the sample is obtained step by step.

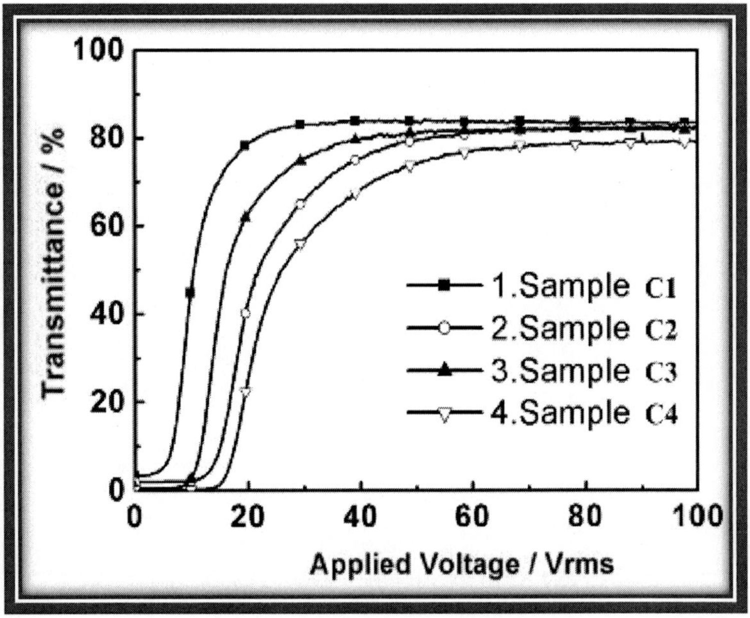

Figure 3-9 shows transmittance vs applied voltage of sample C1-C4

(Figure 3-9) Show transmittance vs applied voltage of sample C1-C4. The sample C1-C4 the transmittance of PDLC film as a function of an applied voltage. Expansion of applied voltage and the transmittance of the sample occur at 80%-83%. Sample C1-C4 shows the complete saturation at 100 volts. As a result, the transmittance is achieved when the applied voltage is increased. The transmittance of the sample was obtained step by step.

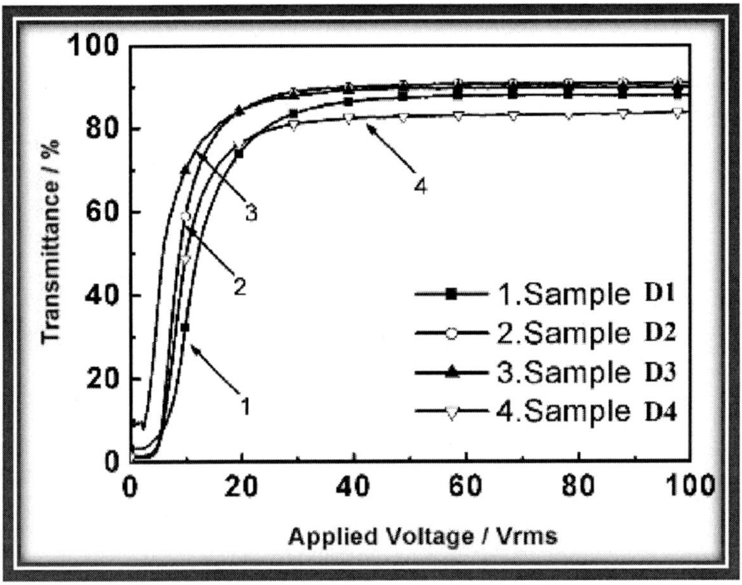

Figure 3-10 shows transmittance vs applied voltage of sample D1-D4

The sample D1-D4 in (Figure 4-10) shows the transmittance of PDLC film as a function of an applied voltage. Expansion of applied voltage and the transmittance of the sample occur at 80%-90%. Sample D1-D4 shows the complete saturation at 100 volts. As a result, the transmittance is achieved when the applied voltage is increased. The transmittance of the sample was obtained step by step.

3.3 The electro-optical properties of PDLC with silver nanoparticles

Performed morphological and electro-optical parameters were studied of UV-cure PDLC. The electro-optical parameter was transmitted light intensity, incident light, droplet number intensity, cross-section, and thickness of PDLC. They were studied along with morphological parameters of liquid crystal droplet diameter, volume fraction, and number intensity studied by SEM. At low curing temperatures, the morphology was increased electro-optical high curing temperatures the morphology decreased and the electro-optical properties (Hakemi et al., 2020).

The author reported highly stretchable hybrid electrodes containing smart windows. The PDLC containing Ag (nanowire) was mixed with the substrate to make PDLC electrode-based smart windows at on-state transmittance of 56% at 80 volts and stretched up to 30%. This PDLC contains high mechanical properties and is used instead of conventional ITO electrodes (Jahangiri & Hamidi, 2016). The paper was based on oxygen-based heterocyclic acrylate monomers with a uniform network morphology and good electro-optical properties were achieved by using oxygen hetero-atom in TFA monomers giving high contrast ratio, less driving voltage, and short response time. Also, polymer film width decreased via 20-18 μm with excellent electro-optical properties were obtained (Saeed et al., 2021).

Performed blue phase liquid crystals were combined with spherical and rod-shaped colloidal semiconductor nanocrystals. The core-shell quantum dots (Qds) and CdSe quantum dots were used as capping agents. The calamitic chiral liquid crystalline compounds with specific quantities were sonicated with Qds and qrs. The POM and differential scanning calorimetry (DSC) were used for optical and thermal analysis. Blue phase can be used for high-quality display and laser application (Khan & Ramarao, 2021).

Prepared PDLC was made up of the azo dye methyl red by mixing a NOA65 polymer and liquid crystal. The dielectric proportion of azo dye methyl red (MR) injected PDLC was studied by the dielectric spectroscopy technique (DST). The MR has a great effect on relaxation time, threshold voltage, conductivity, and dielectric opposed isotropy, relaxation of many cycles and dielectric constant parameters were investigated by experimental results (Önsal, Kocakülah, Kahyaoğlu, & Köysal, 2019).

The work on polymer PDLC glazing is used to study thermal and optical properties. The PDLC film is sandwiched between low iron glasses and attached to the test cell that was placed under an indoor solar simulator. Optical properties of PDLC give good variable transparency at on-state and off-state. PDLC can control the exposure to UV. The transparent state (52.97c and translucent 49.95c) highest temperatures achieved for PDLC glazing that can be used to control UV exposure and replace the old curtains trend in our system (Hemaida, Ghosh, Sundaram, & Mallick, 2020).

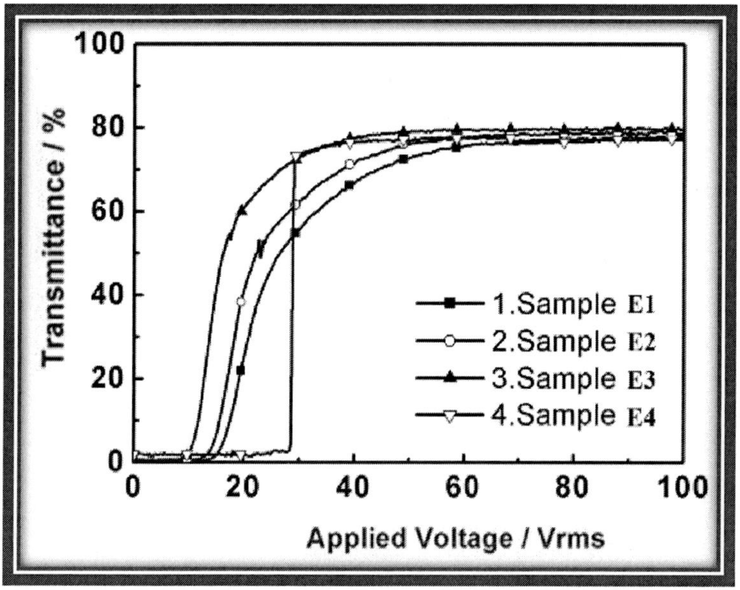

Figure 3-11 shows transmittance vs applied voltage of sample E1-E4

The sample E1-E4 in (Figure 3-11) shows the transmittance of PDLC film as a function of an applied voltage. Expansion of applied voltage and the transmittance of the sample occur at 75%-80%. Samples E1, E2, and E3 show complete saturation at 100 volts. The sample E4 shows the straight peak at 28 volts and saturation at 75%. This may be due to the silver nanoparticle concentration in sample E4. The result indicates transmittance is achieved when the applied voltage is increased. The transmittance of the sample was obtained step by step.

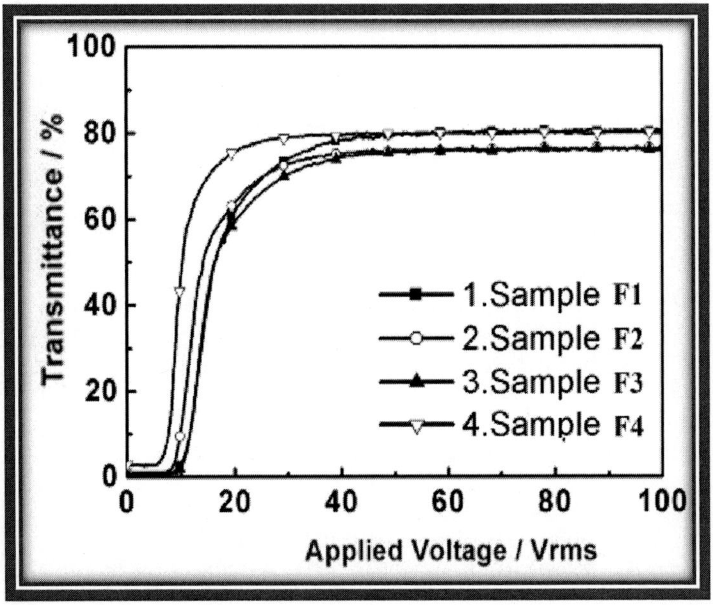

Figure 3-12 shows transmittance vs applied voltage of sample F1-F4

The sample F1-F4 in (Figure 3-12) shows the transmittance of PDLC film as a function of an applied voltage. Expansion of applied voltage and the transmittance of the sample occur at 75%-80%. Sample F1-F4 shows the complete saturation at 100 volts. As a result, the transmittance is achieved when the applied voltage is increased. The transmittance of the sample was obtained step by step.

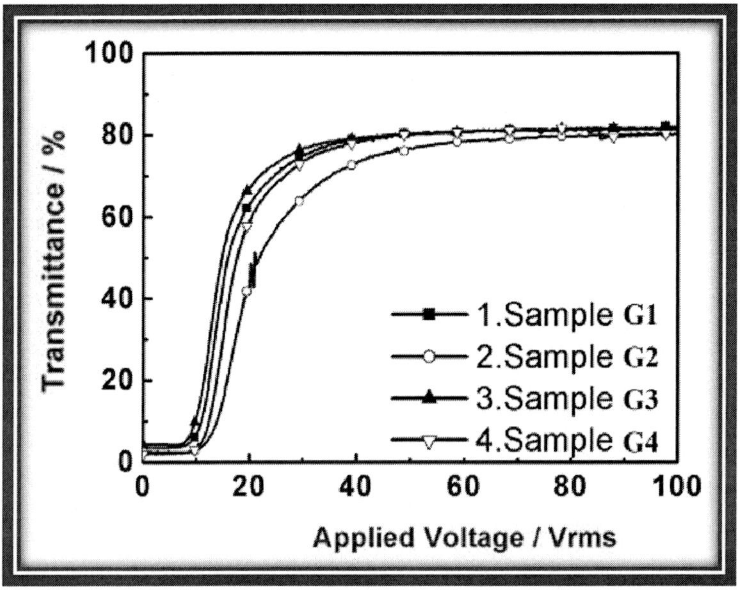

Figure 3-13 shows transmittance vs applied voltage of sample G1-G4

The sample G1-G4 in (Figure 3-13) shows the transmittance of PDLC film as a function of an applied voltage. Expansion of applied voltage and the transmittance of the sample occur at 80%-85%. Sample G1-G4 shows the complete saturation at 100 volts. A result transmittance is achieved when the applied voltage is increased. The transmittance of the sample was obtained step by step.

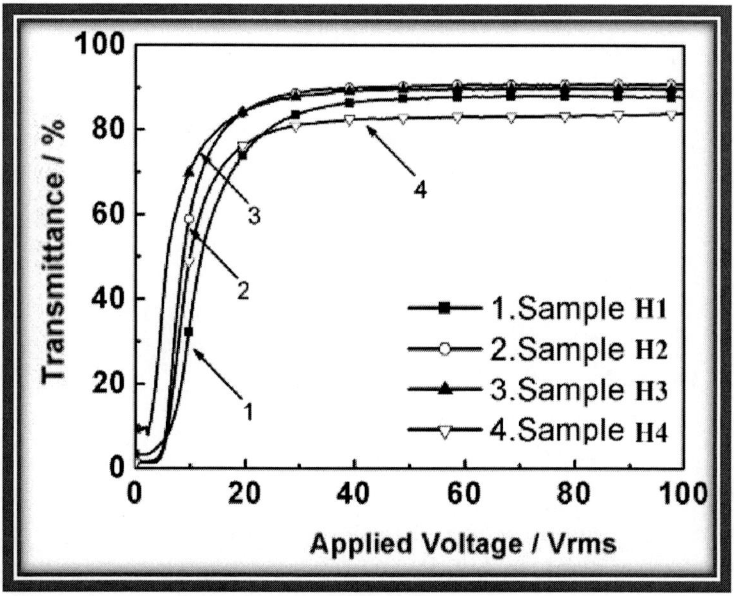

Figure 3-14 shows transmittance vs applied voltage of sample H1-H4

The sample H1-H4 in (Figure 3-14) shows the transmittance of PDLC film as a function of an applied voltage. Expansion of applied voltage and the transmittance of the sample occur at 80%-90%. Sample H1-H4 shows the complete saturation at 100 volts. As a result, the transmittance is achieved when the applied voltage is increased. Transmittance of the sample was obtained step by step.

3.4 Threshold voltage and saturation voltage

The morphology and electro-optical characteristics of the PDLC film strongly depend on the size of the liquid crystal droplet. PDLC film was synthesized by decreasing the liquid crystal content. Hot curable epoxy monomers have been used with different compositions, temperatures, and weight ratios by the heat-curing process (M Ellahi, Gao, & Rafique, 2013).

The low proportions of di and tri epoxy monomers have a great influence on the liquid crystal content of the PDLC film. The functionality and structure of PDLC are influenced by small ratios and by the composition of epoxy and trifunctional monomers. By increasing, the contrast ratio, response time, threshold voltage, the saturation voltage, the morphology of the PDLC is affected (M Ellahi et al., 2015). The morphology and electro-optical properties of crosslinked PDLC are studied by phase separation induced by polymerization. (PIPS) method. Worked on a virgin crosslinked polymer network of PDLCs obtained by increasing the flexible chain length (short and long) and the alkoxy chain length monomers. The more flexible the monomer, the more polymerization of the crosslinked film is caused. The flexible monomers influence the polymerization of the isotopic crosslinked network of PDLC. (M Ellahi, Liu, et al., 2013).

Studied Epoxy monomers that contain an alkyl chain, short flexible chain, and inflexible chain structures reveal that at elevated temperatures short flexible chain, the trifunctional monomer, and the inflexible chain fragment are prepared to provide a polymer matrix with a lower liquid crystal content affects the arrangement and electro-optical properties of the PDLC film. The Multifunctional epoxy monomers with increased temperature increase the curing process (M Ellahi et al., 2014).

The threshold voltage is a physical parameter of a metal-oxide-semiconductor field-effect transistor (MOSFET) where the minimum voltage is required for the device to turn on. The saturation voltage lies in the forward-biased junction is the voltage between emitter and collector terminals.

By decreasing the EGDE monomer the Vth and Vsat increased. The LC content and epoxy monomer affect the PDLC film operating voltage by decreasing the LC droplet size the Vth and Vsat increase. Decreased size of LC size causes the increase in collaboration of polymer matrix and LC molecule. The direction of the applied electric field in a small amount of LC droplets is difficult to orient into the direction of the applied field was done by rearranging the epoxy monomer concentration to operate the PDLC with a specific LC amount.(Mormile et al., 2000)

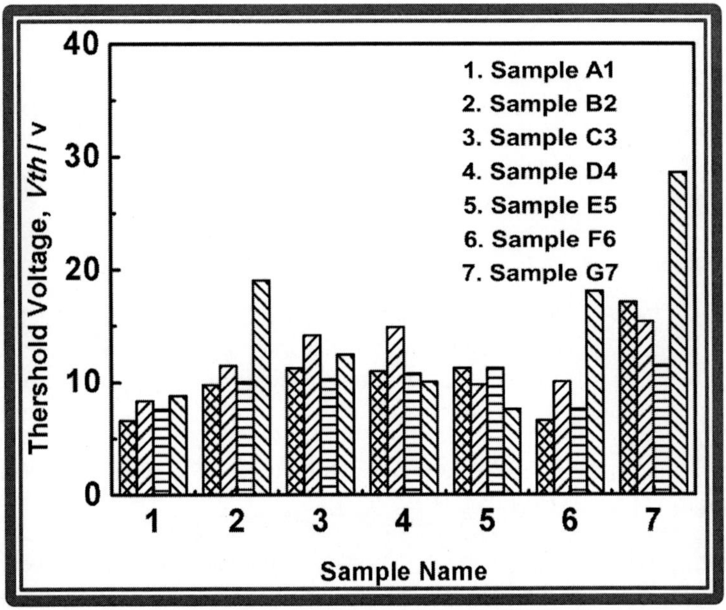

Figure 3-15 shows threshold voltage of Samples A1-G7

Sample 1

The samples A1 and A2 in (Figure 3-15) give threshold voltages of approximately 6.5v and 8v. The sample A1 and A2 Vth are mainly by the EGDE monomer concentration which gives less Vth for A1 while high for the A2 sample has less concentration of EGDE monomer. Sample A3 and A4 give threshold voltage at 7.5v and 8.5v.

Sample 2

Sample B1 and B2 give threshold voltage at 10v and 11.5v. The samples B3 and B4 give threshold voltage at 10v and 19v.

Sample 3

Sample C1 and C2 give threshold voltage at 11v and 14v meanwhile samples C3 and C4 give threshold voltage at 10v and 12.5v.

Sample 4

Sample D1 and D2 give threshold voltage at 12v and 15v and samples D3 and D4 give threshold voltage at 12v and 10v.

Sample 5

Sample E1 and E2 give threshold voltage at 11v and 9v.sample E3 and E4 give threshold voltage at 11v and 7v.

Sample 6

Sample F1 and F2 give threshold voltage at 6v and 10v. While F3 and F4 give threshold voltage at 11v and 18v.

Sample 7

Sample G1 and G2 give threshold voltage at 17v and 15v. Samples G3 and G4 give threshold voltage at 11v and 28v.

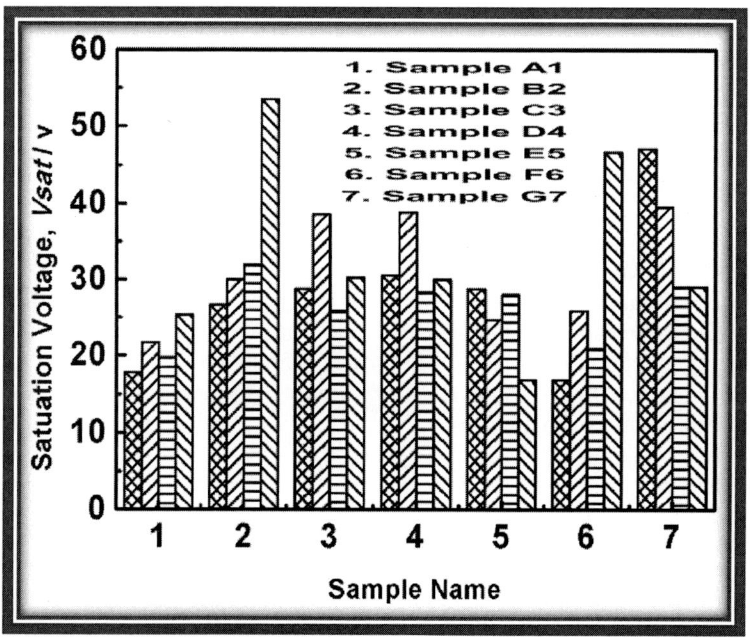

Figure 3-16 Shows saturation voltage of Samples A1-G7

They collected ene-monomer dispersed in the polymer material the variable weight ratios and the liquid crystal content of it. Chain multiplication of the liquid crystal ene-monomer in the polymer matrix becomes functional at lower voltages in Controllable monomer weight ratios. In photoinitiated polymerization reaction. The one-monomer increases the functionality and physical properties (Zhong, Mandle, Goodby, Zhang, & Zhang, 2019). Good electro-optical properties of several methods are available for the synthesis of PDLC films. Few of them have been used solvent-induced phase separation, Polycondensation, radical polymerization, polymer-induced phase separation method (PIPS), emulsion, and temperature-induced phase separation. Composites have been used in different ways where few composites were miscellaneous, composites with a binder, and composites with very little binder have smectic and nematic liquid crystals have been used. Also studied the network of the polymer nanocomposites dispersed in the liquid crystal (Bouteiller & Barny, 1996).

3.5 Saturation voltage

Studied Doping nanocomposites as a whole new technology in the field of smart glass. In the polymer network nanocomposites doped in dispersed liquid crystals, PDLC doped with nanoparticles with electro-optical properties was highly dependent on material anisotropy, the size and shape of the liquid crystal, high contrast ratio, response time, voltages, temperature, and refractive indices of liquid crystal and matrix polymer to form a transparent material(Ahmad et al., 2011).

Studied UV irradiation, thiol vinyl with and without a photoinitiator (PI) was prepared and analyzed with a spectrophotometer thionyl PDLC film was synthesized without a photoinitiator (PI) because it degraded the film. Different weight ratios of polymer and liquid crystal were used with varying stoichiometric ratios of thiol to vinyl. It was possible to study the stoichiometric ratios using FTIR absorbance spectroscopy (Lee, Kim, Ahmad, Jamil, & Jeon, 2014).

Sample 1

The samples A1 and A2 in (Figure 3-16) give saturation voltage achieved approximately at 18v and 21v. Saturation voltage Vsat is also affected by the EGDE monomer concentration it gives good Vsat voltage by adjusting the concentration of EGDE monomer. Sample A3 and A4 give saturation voltage at 19.5v and 25v.

Sample 2

Sample B1 and B2 give saturation voltage at 28v and 30v. The samples B3 and B4 give saturation voltage at 31v and 54v.

Sample 3

Sample C1 and C2 give saturation voltage at 29v and 39v meanwhile samples C3 and C4 give saturation voltage at 28v and 30v.

Sample 4

Sample D1 and D2 give saturation voltage at 30v and 39v. Samples D3 and D4 give saturation voltage at 29v and 30v.

Sample 5

Sample E1 and E2 give saturation voltage at 29v and 27v. Sample E3 and E4 give saturation voltage at 28v and 18v.

Sample 6

Sample F1 and F2 give saturation voltage at 18v and 26v. While F3 and F4 give saturation voltage at 20v and 46v.

Sample 7

Sample G1 and G2 give saturation voltage at 46.5v and 28v. Sample G3 and G4 give saturation voltage at 27v and 27v.

4.6 Contrast ratio

The contrast ratio is an important standard in the electro-optical properties of PDLC film. The contrast ratio is the ratio between T_{on}/T_{off}.

$$CR=T_{on}/T_{off} \tag{3-1}$$

The contrast ratio is strongly influenced by Ton because the contrast ratio and T_{off} are inversely proportional to each other. By decreasing liquid crystal size contrast ratio increases. In samples 1, 2, 3, 4, 5, and 7 (Figure 3-17) the contrast ratio shows variation increases then decreases. Sample 6 shows a decrease to an increase in the ratio.

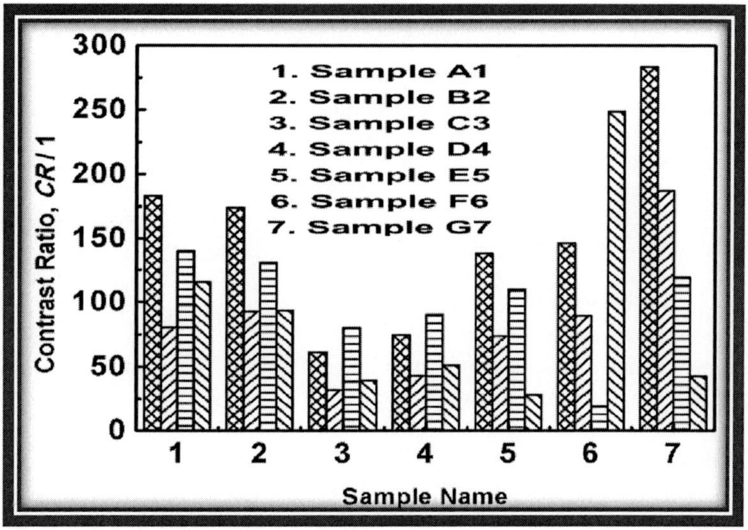

Figure 3-17 shows the Contrast ratio of sample A1-G7

3.7 Rise time and decay time

The rise time and the decay time are the differences between the transparent (rise time tr) and to non-transparent state (decay time td). The tr and td are dependent on the force that opposes such orientation forces like electric field strength molecular orientation and other parameters. This is an important part of PDLC properties for the liquid crystal to align along the direction of the applied electric field and back to its original form upon removing the field. By turning on it the maximum time to cause transmittance from 10-90 % upon turn-off, is 90-10%.

In samples 1,3,4,6 and 7 in (Figure 3-18) the rise time is increasing and then decreasing. While the samples 2 and 5 give a decrease to increase the time. The decay time in (Figure 3-19) samples 1 and 6 is decreasing and remains unchanged in sample 3. While samples 2, 4, 5 and 7 are increasing. The overall transparency is increasing from 10-90% the opaque state decreases from 90-10%.

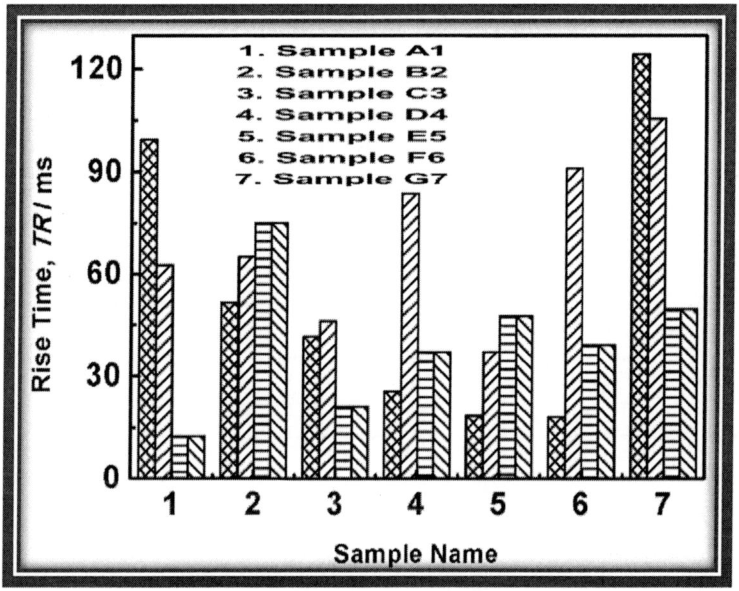

Figure 3-18 shows the Rise time of sample A1-G7

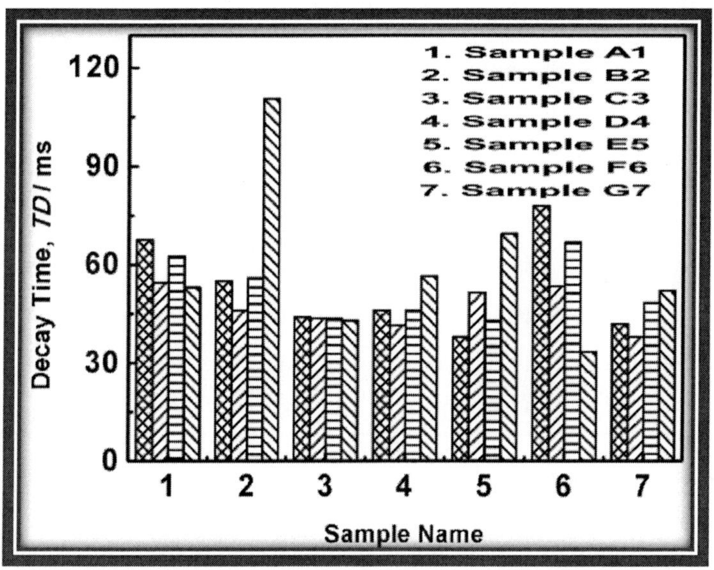

Figure 3-19 show the Decay time of Sample A1-G7

3.8 Refractive index measurement

The refractive index of a matrix was measured using an Abbe Refractometer (WYA-2WAJ, Hangzhou Kebo Instrument Co., Ltd.) at room temperature. The following table monomer mixture has been used for refractive index measurement. Shows the monomer mixture of sample A1-D4 in Table 3-1, 3-2, 3-3 and 3-4.

Table 3-1 shows the monomer mixture of sample A1-A4

sample	Monomers Mixtures (60 wt%)		Curing Temperature (K)
	MM1/MM2 (wt: %)	NLC-1717 / wt: %	
A1	50/10	40	343.15
A2	45/15	40	343.15
A3	40/20	40	343.15
A4	35/25	40	343.15

Monomer mixture [a] (MM1) EGDE / BADGE = 4/2 (molar ratio);
Monomer mixture [b] (MM2) EGDE / BADGE / TETA =3/2/1 (molar ratio)

Table 3-2 shows the monomer mixture of sample B1-B4

Sample	Monomers Mixtures (60 wt%)		Curing Temperature (K)
	MM1/MM2 (wt: %)	NLC-1717 / wt: %	
B1	50/10	40	343.15
B2	45/15	40	343.15
B3	40/20	40	343.15
B4	35/25	40	343.15

Monomer mixture [a] (MM1) EGDE / BADGE = 4/2 (molar ratio);
Monomer mixture [b] (MM2) EGDE / BADGE / DETA =2/2/3 (molar ratio)

Table 3-3 shows the monomer mixture of sample C1-C4

Sample	Monomers Mixtures (60 wt%)		Curing Temperature (K)
	MM1/MM2 (wt: %)	NLC-1717 / wt: %	
C1	50/10	40	343.15
C2	45/15	40	343.15
C3	40/20	40	343.15
C4	35/25	40	343.15

Monomer mixture [a] (MM1) EGDE / BADGE = 4/2 (molar ratio);
Monomer mixture [b] (MM2) EGDE / BADGE /PERTEMP =2/2/3 (molar ratio)

Table 3-4 shows the monomer mixture of sample D1-D4

Sample	Monomers Mixtures (60 wt%)		Curing Temperature (K)
	MM1/MM2 (wt: %)	NLC-1717 / wt: %	
D1	50/10	40	343.15
D2	45/15	40	343.15
D3	40/20	40	343.15
D4	35/25	40	343.15

Monomer mixture [a] (MM1) EGDE / NGDE= 4/2 (molar ratio);
Monomer mixture [b] (MM2) EGDE / NGDE / PERTEMP =2/2/3 (molar ratio)

3.8.1 Refractive index for sample A1-A4

Table 3-5 shows refractive indices (np) of domain-size smart glass. The refractive index is the changing path of a light ray from one intermediate to another. The amount of mismatching of the refractive index decreased first and then increased from sample A1 to A4. The smaller the mismatching degree of the refractive index of the polymer matrix and LC droplet the higher the T_{off} of PDLC. The increase in T_{off} gives the decreased contrast ratio because the contrast ratio and Toff are inversely proportional to each other.

Table 3-5 shows the refractive index of sample A1-A4

Sample	Refractive index			
	Mixture[a]	n_p[b]	n_o[c]	n_p-n_o
A1	1.5093	1.4961	1.532	-0.0359
A2	1.4711	1.4475	1.532	-0.0845
A3	1.4628	1.5466	1.532	0.0146
A4	1.4407	1.4947	1.532	0.0304

Conditions

[a] The refractive index of the mixture before curing.

[b] The refractive index of polymer matrix without dissolved LC material.

[c] The ordinary refractive index of NLC-1717 was used.

3.8.2 Refractive index for sample B1-B4

Table 3-6 shows refractive indices (np) of domain-size smart glass. The refractive index is the changing path of a light ray from one intermediate to another. The amount of mismatching of the refractive index decreased first and then increased from sample B1 to B4. The smaller the mismatching degree of the refractive index of the polymer matrix and LC droplet the higher the T_{off} of PDLC. The increase in T_{off} gives the decreased contrast ratio because the contrast ratio and T_{off} are inversely proportional to each other.

Table 3-6 shows the refractive index of sample B1-B4

Sample	Refractive index			
	Mixture[a]	n_p[b]	n_o[c]	n_p-n_o
B1	1.5053	1.4821	1.532	-0.0499
B2	1.4978	1.4456	1.532	-0.0864
B3	1.3901	1.5510	1.532	0.019
B4	1.3709	1.5928	1.532	0.0608

Conditions

[a] The refractive index of the mixture before curing.

[b] The refractive index of polymer matrix without dissolved LC material.

[c] The ordinary refractive index of NLC-1717 was used.

3.8.3 Refractive index for sample C1-C4

Table 3-7 shows refractive indices (np) of domain-size smart glass. The refractive index is the changing path of a light ray from one intermediate to another. The amount of mismatching of the refractive index decreased first and then increased from sample D1 to D4. The higher the mismatching degree of the refractive index of the polymer matrix and LC droplet the lower the Toff of PDLC. The reduced Toff gives the increased contrast ratio because the contrast ratio and Toff are inversely proportional to each other.

Table 3-7 shows the refractive index of sample C1-C4

Sample	Refractive index			
	Mixture[a]	n_p[b]	n_o[c]	n_p-n_o
C1	1.6782	1.4936	1.532	-0.0384
C2	1.5834	1.4583	1.532	-0.0737

C3	**1.4721**	**1.6083**	**1.532**	**0.0763**
C4	**1.4374**	**1.5964**	**1.532**	**0.0644**

Conditions
[a] The refractive index of the mixture before curing.
[b] The refractive index of polymer matrix without dissolved LC material.
[c] The ordinary refractive index of NLC-1717 was used.

3.8.4 Refractive index for sample D1-D4

Table 3-8 shows refractive indices (np) of domain-size smart glass. The refractive index is the changing path of a light ray from one intermediate to another. The amount of mismatching of the refractive index increased first and then decreased from sample D1 to D4. The smaller the mismatching degree of the refractive index of the polymer matrix and LC droplet the higher the T_{off} of PDLC. The increase in T_{off} gives the decreased contrast ratio because the contrast ratio and T_{off} are inversely proportional to each other.

Table 3-8 shows the refractive index of sample D1-D4

Sample	**Refractive index**			
	Mixture[a]	n_p[b]	n_o[c]	n_p-n_o
D1	**1.5210**	**1.5423**	**1.532**	**0.0103**
D2	**1.5076**	**1.5678**	**1.532**	**0.0358**
D3	**1.4989**	**1.4819**	**1.532**	**-0.0501**
D4	**1.4636**	**1.4428**	**1.532**	**-0.0892**

Conditions
[a] The refractive index of the mixture before curing.
[b] The refractive index of polymer matrix without dissolved LC material.
[c] The ordinary refractive index of NLC-1717 was used.

Conclusion

In conclusion, the performed work showed an effective result in realizing a wise glass thin-film PDLC with high distinction magnitude relation and low driving voltages. The ready three epoxy resin-based PDLC skinny films are made by doped AgNPs. The thin PDLC (smart glass) polymer network was powerfully controlled by the structure of the heat-curing epoxy monomers and ratios. UV-vis spectrum results show silver nanoparticles peaks at 395,398,400 and 405 nm wavelength that is the right peaks for silver nanoparticles.

The electro-optical properties show that the PDLC film Ready with silver nanoparticles system has samples A1-A4, indicating that the driving voltage will increase about the variable magnitude of heat-curable chemical compounds like EGDE, BADE, and TETA. The threshold voltage (Vth), saturation voltage (Vsat) and contrast ratio (CR) Increase whereas the on-state and off-state transmission of PDLC films decreases.

While the electro-optical properties of PDLC film are ready while not silver nanoparticles having sample B1-B4 the driving voltage will increase with variable magnitude relation heat-curable chemical compound EGDE, BADE, and DETA. The threshold voltage (Vth), saturation voltage (Vsat), and metallic element increase, and therefore on-state and off-state transmission of PDLC film decrease, While the electro-optical properties of PDLC film ready while not silver nanoparticles having sample C1-C4 the driving voltage will increase with variable magnitude relation heat-curable chemical compound EGDE, BADE and PERTEMP.

The threshold voltage (Vth), saturation voltage (Vsat) and the metallic element increase, and therefore on-state and off-state transmission of PDLC film decrease. The electro-optical properties of PDLC film ready while not silver nanoparticles having sample D1-D4 the driving voltage will increase with variable magnitude relation heat-curable chemical compound EGDE, NGDE and PERTEMP. The threshold (Vth), saturation voltage (Vsat), and metallic element increase and therefore on-state and off-state transmission of PDLC film decrease.

The electro-optical properties of PDLC film ready with silver nanoparticles having sample E1-E4 the driving voltage decrease with variable magnitude relation heat-curable chemical compound EGDE, BADE, TETA and AgNPs. The threshold voltage (Vth), saturation voltage (Vsat), and metallic element decrease, and therefore on-state and off-state transmission of PDLC film decrease.

Electro-optical properties of PDLC film ready with silver nanoparticles having sample F1-F4 the driving voltage decreases with variable magnitude relation heat-curable chemical compound EGDE, BADE, DETA, and AgNPs. The threshold voltage (Vth), saturation voltage (Vsat) and metallic element increase and therefore on-state and off-state transmission of PDLC film increase because of the DETA hardener, While the electro-optical properties of PDLCs film ready with silver nanoparticles having sample G1-G4 the driving voltage decreases with variable magnitude relation heat-curable chemical compound EGDE, BAGDE, PERTEMP and AgNPs. The threshold voltage (Vth), saturation voltage (Vsat), and metallic element decrease, and therefore on-state and off-state transmission of PDLC film decrease.

Acknowledgements

The authors acknowledge the State Key Laboratory of Advanced Analytical Polymer Laboratory Optoelectronic Materials and Technologies of BUITEMS, Quetta, for the use of experimental equipment. The author is very grateful to International Polymer Industries (Pvt) Ltd., Islamabad, for technical assistance. The author is very grateful to the Dr. Hamdullah Khan and Dr. Shahista Taimur contributed to the establishment of the research design as well as the synthesis protocol to prepare the PDLC thin films.

The electro-optical property was measured using a liquid crystal device parameters tester (LCT-5016C, Changchun Liancheng Instrument Co. Ltd., China) from the People's Republic of China and the characterization of SEM, an Abbe refractometer, and UV-Vis spectra from the Islamic Republic of Pakistan.

Funding

We would like to acknowledge the Higher Education Commission (HEC) of I.R. Pakistan, a project of Technology development fund (TDF02-054) and No. 01/Research Projects/ORIC/ BUITEMS/340, of BUITEMS, & No.9356/Balochistan / NRPU/R&D/HEC.

References

Ahmad, Farzana, Jamil, Muhammad, Lee, Jin Woo, & Jeon, Young Jae. (2015). Magnetically driven vertical alignment of liquid crystals by ferromagnetic particles. *Liquid Crystals, 42*(2), 233-239.

Ahmad, Farzana, Jamil, Muhammad, Lee, Jin Woo, Ri, Yeo Hyo, & Jeon, Young Jae. (2014). Characteristics of di-and tri-block copolymers: polymer disperse liquid crystal display. *Journal of Modern Optics, 61*(12), 1027-1032.

Ahmad, Farzana, Jeon, A-Ri, Jeon, Young Jae, & Jamil, Muhammad. (2021). A novel technique of fabrication of nanoparticle acrylate doped polymer dispersed liquid crystal (PDLC) film. *Journal of Dispersion Science and Technology,42*(1), 1-6.

Doane, JW, Vaz, NA, Wu, B-G, & Žumer, Slobodan. (1986). Field controlled light scattering from nematic microdroplets. *Applied Physics Letters, 48*(4), 269-271.

Drzaic, Paul, & Drzaic, Paul S. (2006). Putting liquid crystal droplets to work: a short history of polymer dispersed liquid crystals. *Liquid crystals, 33*(11-12), 1281-1296.

Ellahi, Mujtaba, Gao, Yanzi, & Rafique, MY. (2013). Influence of enhanced curing temperature of epoxy monomers structure on the electro-optical properties and morphology of polymer-dispersed liquid crystal films. *American Journal of Engineering Research, 2*(3), 01-06.

Ellahi, Mujtaba, Liu, Fang, Song, Ping, Gao, Yanzi, Cao, Hui, Rafique, MY, Yang, Huai. (2013). Influence of the multi-functional epoxy monomers structure on the electro-optical properties and morphology of polymer-dispersed liquid crystal films. *Polymer bulletin, 70*(11), 2967-2980.

Ellahi, Mujtaba, Liu, Fang, Song, Ping, Gao, Yanzi, Rafique, MY, Khan, Dil Faraz, Yang, Huai. (2014). Characterization and morphology of polymer-dispersed liquid crystal films. *Soft Materials, 12*(3), 339-345.

Ellahi, Mujtaba, Nawaz, Muhammad, & Rafique, M Yasir. (2017). Morphological Study of PDLC Films with Refractive Index Using Catalyst. *Pakistan Journal of Engineering and Applied Sciences.21*(6), 1-5.

Gao, Boxuan, De Jong, Ties, Nie, Chuan, Osterodt, Jens, Neyts, Kristiaan, & Beeckman, Jeroen. (2021). Driving issues of large area liquid crystal devices. *Liquid Crystals, 48*(2), 281-294.

Gracia-Pinilla, MiguelÁngel, Ferrer, Domingo, Mejía-Rosales, Sergio, & Pérez-Tijerina, Eduardo. (2009). Size-selected Ag nanoparticles with five-fold symmetry. *Nanoscale research letters, 4*(8), 896-902.

Gui, Kun, Zheng, Jihong, Wang, Kangni, Li, Daoping, & Zhuang, Songlin. (2015). FDTD modelling of silver nanoparticles embedded in phase separation interface of H-PDLC. *Journal of Nanomaterials, 16*(1),87-89.

Hakemi, H, Khenkin, M, Tamari, R, & Gal-Fuss, D. (2020). Transient Phenomenon in Polymer-Dispersed Liquid Crystals. *Recent Progress in Materials, 2*(1), 1-1.

Hecht, Engene, & Zajac, A. (2002). Optics (4th eds):San Francisco, CA.Addison Wesley. Hecht, Engene, & Zajac.

Hemaida, Abdulmohsin, Ghosh, Aritra, Sundaram, Senthilarasu, & Mallick, Tapas K. (2020). Evaluation of thermal performance for a smart switchable adaptive polymer dispersed liquid crystal (PDLC) glazing. *Solar Energy, 19*(5), 185-193.

Kizhakidathazhath, Rijeesh, Nishikawa, Hiroya, Okumura, Yasushi, Higuchi, Hiroki, & Kikuchi, Hirotsugu. (2020). High-performance polymer dispersed liquid crystal enabled by uniquely designed acrylate monomer. *Polymers, 12*(8), 251-60.

Klapetek, Petr, Valtr, Miroslav, Nečas, David, Salyk, Ota, & Dzik, Petr. (2011). Atomic force microscopy analysis of nanoparticles in non-ideal conditions. *Nanoscale research letters, 6*(1), 1-9.

Krishnakant , G. Mishra, & Sureshchandra, J. Gupta. (2011). Influence of ZnO/CdO Nanoparticles on Embedding in Polymer Dispersed Liquid Crystals *Archives of physics research,2*(3), 84-93.

Kumano, Naomi, Seki, Takahiro, Ishii, Masahiko, Nakamura, Hiroshi, Umemura, Tomonari, & Takeoka, Yukikazu. (2011). Multicolor polymer-dispersed liquid crystal. *Advanced Materials, 23*(7), 884-888.

Kumari, Asha, & Sinha, Aloka. (2020). Role of BaTiO3 nanoparticles on electro-optic performance of epoxy-resin-based PDLC devices. *Liquid Crystals*, *48*(1), 23-34.

Lee, Jin Woo, Kim, Joon Ki, Ahmad, Farzana, Jamil, Muhammad, & Jeon, Young Jae. (2014). Properties of thiol–vinyl PDLC films without additional photoinitiator. *Liquid Crystals, 41*(8), 1109-1115.

Li, Ke, Jiang, Haodong, Cheng, Ming, Li, Yifei, Yin, Zhen, Luo, Dan, Liu, Yan Jun. (2020). Controlling morphological and electro-optical properties via the phase separation in polymer/liquid-crystal composite materials. *Liquid Crystals, 47*(2), 238-247.

Li, Lu, Zhang, Yi, Liao, Xia, Hao, Zengheng, Sheng, Xingyue, & Liu, Pan. (2019). Structure and Properties of Poly (Oxypropylene) Diamine Intercalated Montmorillonite/Epoxy Composites. *Journal of Macromolecular Science, Part B, 58*(11), 877-889.

Liu, Yourong, Zheng, Jihong, Shen, Tong, Wang, Kangni, & Zhuang, Songlin. (2019). Diffusion kinetics investigations of Nano Ag-doped holographic polymer dispersed liquid crystal gratings. *Liquid Crystals, 46*(12), 1852-1860.

Masutani, A, Hollfelder, N, Pleis, F, Roberts, T, Kilickiran, P, Sakaigawa, A, Nelles, G. (2007). *Polymer dispersed liquid crystal (PDLC) for paper-like display application.* Paper presented at the 14th International Display Workshop (pp.264-268). Sapporo, Japan.

McFarland, Coleen A, Koenig, Jack L, & West, John L. (1993). Influence of the polymer matrix on the bipolar and radial droplet configurations within PDLC films examined by infrared spectroscopy. *Applied spectroscopy, 47*(5), 598-605.

Mehr, Ferdos Parsa, Khanjani, Masoumeh, & Vatani, Parya. (2015). Synthesis of nano-Ag particles using sodium borohydride. *Oriental Journal of Chemistry, 31*(3), 1831-1833.

Miller, Sanford S, & Mocanu, Petru T. (2000). *Differential subordinations: theory and applications*: (1st eds) England, UK.CRC Press. Miller, Sanford S, & Mocanu, Petru T 480.

Song, Ping, Cao, Hui, Wang, Feifei, Ellahi, Mujtaba, & Yang, Huai. (2012). Study of polymer-dispersed liquid crystal systems using epoxies/acrylates as hybrid polymer matrix components. *Liquid Crystals, 39*(8), 903-909.

Song, Ping, Cao, Hui, Wang, Feifei, Liu, Fang, Wang, Jingjing, Ellahi, Mujtaba, Yang, Huai. (2012). The UV polymerisation temperature dependence of polymer-dispersed liquid crystals based on epoxies/acrylates hybrid polymer matrix components. *Liquid Crystals, 39*(9), 1131-1140.

Sparavigna, Amelia Carolina. (2019). Polymer dispersed liquid crystals and related electro-optical devices-An overview .HAL Id: Hal -02265643, version 1.

Tanaka, Yoshio. (2018). Synthesis and characteristics of epoxides, *Epoxy resins: chemistry and technology* (2nd ed ., pp. 9-283).United kingdom, England:Routelege taylor and francis England.

Tao, Lei, Sun, Zeyu, Min, Wei, Ou, Hanwen, Qi, Liangliang, & Yu, Muhuo. (2020). Improving the toughness of thermosetting epoxy resins via blending triblock copolymers. *RSC Advances, 10*(3), 1603-1612.

Unnikrishnan, KP, & Thachil, Eby Thomas. (2006). Toughening of epoxy resins. *Designed monomers and polymers, 9*(2), 129-152.

Vaz, Nuno A, & Montgomery Jr, G Paul. (1987). Refractive indices of polymer-dispersed liquid-crystal film materials: Epoxy-based systems. *Journal of applied physics, 62*(8), 3161-3172.

Wang, Hongshui, Qiao, Xueliang, Chen, Jianguo, & Ding, Shiyuan. (2005). Preparation of silver nanoparticles by chemical reduction method. *Colloids and Surfaces A: Physicochemical and Engineering Aspects, 256*(2-3), 111-115.

Wang, Huihui, Gong, He, Song, Ping, Sun, Jian, Guo, Shumeng, Cao, Hui, Yang, Huai. (2015). Reverse-mode polymer dispersed liquid crystal films prepared by patterned polymer walls. *Liquid Crystals, 42*(9), 1320-1328.

Yan, Xudong, Zhou, Yong, Liu, Wei, Liu, Sunqian, Hu, Xiaowen, Zhao, Wei,Yuan, Dong. (2020). Effects of silver nanoparticle doping on the electro-optical properties of polymer stabilized liquid crystal devices. *Liquid Crystals, 47*(8), 1131-1138.

Zhang, Cuihong, Ge, Yuan, Huo, Xiaoping, Xue, Jing, Li, Kexuan, Zhang, Yongming, & Miao, Zongcheng. (2020). Studies on electro-optical properties of polymer matrix/LC/ITO nanoparticles composites. *Polymers for Advanced Technologies, 31*(3), 544-552.

CHAPTER 4

Abstract

Polymeric dispersion compounds (PDLCs) have been generated by using the starting-to-turn emulsification polymerization-induced phase separation (PIPS) method and a low birefringent -longitudinal serve to maintain nematic liquid crystal. The effect of penta functionality was examined and compared with both of the doped samples of PDLC flexible thin films. With long flexibility longer chain ethylene glycol diglycidyl ether (EDGE) polymer dispersal on electrical (E-O) properties of PDLC, Neo pentyl glycol diglycidyl diethyl (NGDE) is currently offered for sale. The input impedance rises by 62%, overall reaction time improves by 80% in PDLC produced with a 20% weight per cent ratio of NLC. The ON sticky glue agent hardener, Triethylenetetramine (TETA), was found to improve the optical clarity and transparency of the picture. Vth and Vsat grow with decreasing plastic mesh size, tonnage increases with decreasing polymer mesh size, toff increases with decreasing copolymer mesh size, and off-state penetration rises with decreasing copolymer mesh size. Because of the long cellulose chains of the merge hardliner TETA and the significant increase in the copolymer length between two cross-linking agents, the effect of the PDLC materials on house music rental homes is more evident than it is for flexible thin film results. However, it is also important to consider the differences in viscosity, heat stability, and monomer methyl (-CH3) chain length among the different cross-linking agents. These displays will act as a protective barrier between the health of the COVID-19 sufferer and the physician. The Isolation chambers, institutions, smart headgear, and smart cities all help to close social gaps.

4.1 Connectivity structure in polymers

Table 4.1 the compositions of sample A-G, Curing time 7 hours and at 343.15 K

Sample	Monomers Mixtures (80 wt. %) MM1[a]/MM2[b]	SLC1717/wt. %
A-1	**70/10**	**20**
B-2	**65/15**	**20**
C-3	**60/20**	**20**
D-4	**55/25**	**20**
E-5	**50/30**	**20**
F-6	**55/35**	**20**
G-7	**50/40**	**20**

The current work set out a technique to create thin polymer-dispersed liquid crystal (PDLC) films for smart glasses by utilizing an epoxy resin monomer framework, hardeners, and a nematic liquid crystal system. Optical and electro-optical display panel effectiveness is directly influenced by the approaches employed to enhance the stability and properties of smart PDLC films. In the present study, the polymerization-induced phase separation heat curing method is set up using epoxy resin with tetra-functional pentaerythritol tetra glycidyl ether, tri-functional trimethylolpropane triglycidyl ether, and thermosetting resin 3-(diethyl amino)propyl amine as the primary functional heat curing agent and hardener. Using a scanning electron microscope, UV-Vis-NIR spectrophotometer, liquid crystal device parameter tester, and the sheets of the created smart glass panels. The interaction between the monomer and the principal functional hardener, 3-(diethyl amino) propyl amine, was verified using the liquid crystal device approach and scanning electron microscopy. Tests were carried out on epoxy resins that were cured using 3- (diethyl amino) propyl amine to ascertain the curing temperature, active diluent ratio, and epoxy monomer feed ratio. The primary variable reducing the domain size of PDLC thin films in all samples was also shown to be the methyl groups of pentaerythritol tetraglycidyl ether, 3-(diethyl amino)propyl amine, and trimethylolpropane triglycidyl ether. This smart glass technology will separate the COVID-19 patient and the physician, and will also be used in hospitals, affected patient isolation chambers, healthcare centre buildings, and smart helmets to facilitate social distancing and environment-friendly electronics [28].

The findings on the electro-optical characteristics of thin polymer films containing CdSe/ZnS quantum dots and switchable periodic shapes holographically generated in polymer-dispersed liquid crystals (PDLCs) are compiled in this paper. We specifically address the following topics: (1) the influence of gamma-radiation and microstructure on the electro-optical properties of PDLCs; (2) the switching behaviour of periodic patterns based on PDLCs as well as complementary "reverse-mode" structures; and (3) the effects of gamma-ray and 532-nm laser irradiations concurrently on the photoluminescence properties of thin films that included CdSe/ZnS quantum dots [29].

Through a solvent-induced phase separation technique, polymer-dispersed liquid crystal (PDLC) films (5CB/PMMA, 60/40) with varying droplet sizes have been produced at various N2 flow rates. They were the effects of droplet size on the conductance, dielectric loss, dielectric constant, and the electric field created in a droplet, as well as the thermal transitions of the LC. By combining the film's dielectric constant with the value anticipated by Boettcher's mixture formula, the configuration of the LC in the film containing tiny droplets can be determined. Furthermore, the impact of the size of droplets on the PDLC film's electro-optical response was examined. Under different AC frequencies, changes in the film's conductance and dielectric constant were examined to define the LC molecules' droplet-based polarization mechanism [30].

This study comprised the manufacturing of polymer-induced phase-separated low molecular weight thermotropic liquid crystal (LC) and high molecular weight polymer, leading to polymer dispersed liquid crystal (PDLC) films. LC was coated with dichroic azo dye to increase the optical efficiency of the PDLC films. The characteristics of the constituent parts and the working principles of the dye-doped PDLC (DPDLC) films, various kinds of characterization approaches were utilized. The morphological investigations conducted by scanning electron microscopy as well as polarized optical microscopy revealed that the configuration of the LC droplets varied depending on the voltage and concentration of the dye. The results of the electro-optical investigation demonstrated that the DPDLC film performed exceptionally well at low dye concentrations (0.007%), exhibiting outstanding contrast ratios (241) and transmittance differences ($\Delta T = 98.57\%$).

The minimum hysteresis impact, low saturation (Vsat = 1.45 V/μm) and threshold (Vth = 0.85 V/μm) voltages. The study investigated the impact of temperature on the electro-optical characteristics of DPDLC films. UV-visible spectrophotometry was utilized to perform absorbance analyses to supplement the transmittance analyses. The relaxation time and frequency of DPDLC films were measured using dielectric relaxation spectroscopy. The purpose of Debye and Cole-Cole's modelling was to comprehend the nature of the relaxing process. The Debye-type relaxation process of all the DPDLC composite films was validated by their distribution parameter (α) having values of zero [31].

Fig 4.1.2 of PDLC instances A-1 through G-7, the bonding strength of the zeolite decreases and indeed the wire diameter of the zeolite increases when the amount of NGDE and EGDE in the extended flexible chain rises. The following three elements can be utilized to explain why a rise in NGDE and EGDE contents increases the copolymer filter One is that the cross-linking point of the polymer network falls as NGDE and EGDE molecular chains lengthen, and consequently, its mesh has grown; the third, because the mass ratio of NGDE and EGDE with short molecular chains changes, the viscosity and melting point of NGDE and EGDE are higher, respectively, and the number of moles in the system decreases as the content of TETA increases. This results in a decrease in the degree of cross-linking of the polymerization, which causes the polymer mesh to increase. It was causing the diffusion of liquid crystal molecules in the mixture solution to be more challenging, which in turn caused the polymer mesh to grow and produce the minor polymer.

Fig 4.1 Micrometer electrochromic particles make up PDLCs.

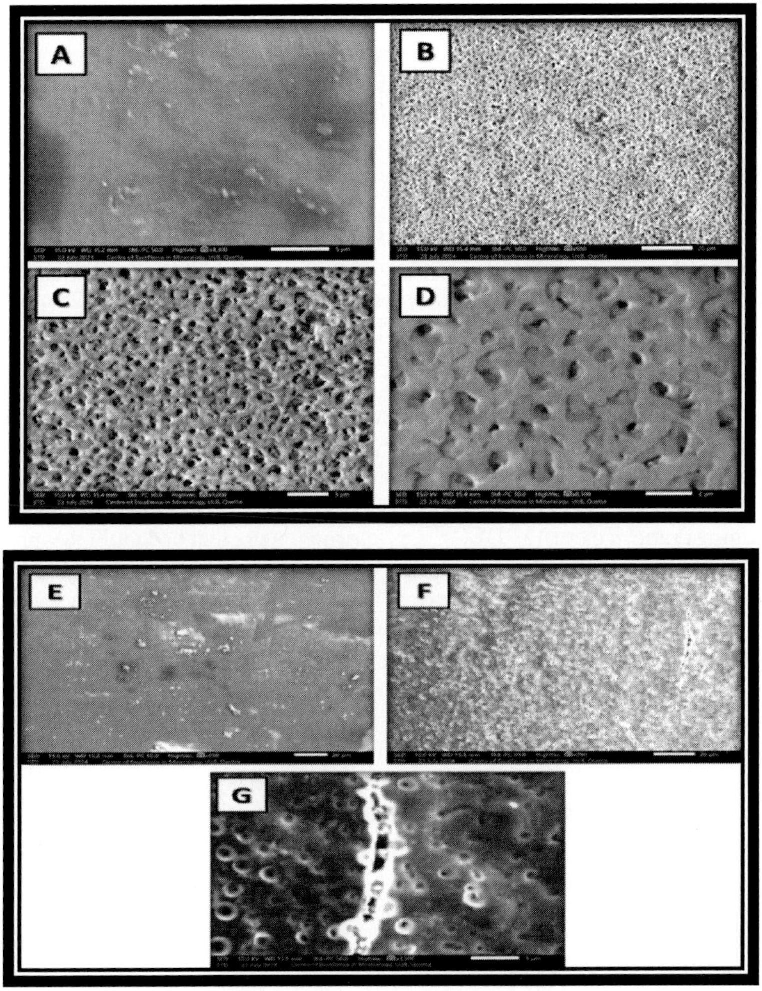

Fig 4.2 SEM micrographs of the polymer networks of the samples A-G

Utilizing cholesteryl acetate liquid crystal and chitosan, a collection of five ecologically friendly PDLC composite films has been developed and evaluated. The resulting composites' form and structure were assessed using scanning electron microscopy, FTIR and RAMAN spectroscopy, and polarized optical microscopy. Their thermotropic behaviour was investigated using differential scanning calorimetry and polarized optical microscopy. Furthermore, measurements and calculations were made for the surface free energy part and contact angle. The results showed that chitosan can cause the cholesteryl acetate liquid crystals to form as radial drops of water, disrupting the supramolecular chiral ordering of the element due to the existence of hydroxyl, amine, and amide groups in its structure. The mass ratio between the polymer and the liquid crystal determined the droplets' shape and density. Micrometric droplets with a diameter of 5 μm to 8 μm frequently comprise the most appropriate sizes for optoelectronic applications. By altering the mass ratio of the constituents, PDLC composites with varying hydrophobicities have been created as a consequence of their surface's morphological microstructure and chemical composition [32].

The remarkable electro-optical properties of polymer dispersed liquid crystals (PDLCs) have been attributed to their distinct phase-separated microstructure. Nevertheless, the low contrast ratio and high driving voltage limit the scope of their use. The building components have been optimized to lower the driving voltage without compromising other attributes like the contrast ratio. To create PDLCs, acrylate monomers that had multiple steric end groups were added, and the effects of these monomers on the microstructure and electro-optical characteristics were examined. All samples showed bipolar droplet configuration, suggesting tangential anchoring circumstances. It was found that the electro-optical characteristics and polymer matrix morphology might be influenced by the steric end group. It was possible to improve the contrast ratio and the driving force by carefully manipulating the steric group and doping concentration. There could be a voltage drop. In addition, hysteresis could be reduced. The investigation enhances the possible applications of PDLCs by providing more possibilities for their creation through a material perspective [33].

Due to its unique characteristics, polymer-dispersed liquid crystal (PDLC) film is an active smart film that has worked its way into the market. These beneficial benefits include the capacity to adjust the tint, which protects inhabitants of the structure from UV radiation from the sun, enhances energy-saving features, and increases cost-efficiency. Even while PDLC films show promise in several applications, their performance is still unclear. The link between the size effects (film thickness and area) and light intensity, apparent power consumption, UVR, IRR, visual light transmission (VLT), light intensity, and current consumption is particularly poorly known. Consequently, the size consequences of PDLC films have been investigated in this work, along with the impact of area and thickness on current and perceived power consumption as well as the effects of light intensity, UVR, IRR, and VLT. It also successfully verified how the PDLC film's light transmittance was influenced by modifications to the applied voltage. It was successfully demonstrated that a 0.1 mm PDLC film had ideal characteristics and was cheap to purchase. Thus, this study opens the door to a better knowledge of PDLC films (act and sizing impacts) and how to use cost-effective PDLC films for widespread implementation in residential and commercial spaces [34].

Buildings require an extensive amount of energy to provide artificial sunlight and air cooling. The glass technology serves as the primary enclosing element in buildings for decorative and natural daylighting purposes, yet there is significant solar heat gain and loss via the clear glass. The purpose of this research is to investigate the solar optical characteristics and potential savings in costs for air conditioning of several smart PDLC film glasses. The solar optical characteristics of four distinct smart PDLC film glasses white, blue, pink, and yellow both with and without applied voltage circumstances are presented in this work. To determine solar heat gain through smart PDLCs in voltage ON/OFF states, a numerical framework was created. Additionally, a cost analysis was done to calculate the yearly savings on air conditioning.

Every intelligent PDLC film eyewear in voltage ON/OFF circumstances showed a significant decrease in heat gain/loss in buildings of three climate conditions when compared to standard clear glass. The net yearly cost reductions (heating cost + cooling cost) were caused by the smart PDLC film glasses' reduced heat gain/loss. Comparing the white smart PDLC film glass WSPG (V) to the other studied smart glasses in eight orientations of three climatic conditions, it was found to be the most energy-efficient smart glass with the highest annual air-conditioning cost savings ($101.76 in the SE of hot and dry climate), lowest return on investment (12.71 yrs in the SE of hot and dry climate), and adequate daylight factors. The findings support the design and selection of appropriate glass for solar-passive structures that are both energy-efficient and sustainable [35].

According to environmental scientists' scientific predictions regarding energy usage, one of the most serious contemporary global challenges is energy scarcity. Energy-saving solutions are therefore becoming more and more important since they have positive effects on the environment, economic growth, and individual users. To maintain energy availability while reducing detrimental effects on the environment, more scientific studies are being done on improving the efficiency of energy at a slower rate (e.g., thermal pollution, glass house global warming effects). Working towards this vital goal, this study provides a new approach to creating a smart and efficient energy-saving system employing PDLC glass. It can steadily change from a remote control to the user's preferred shades automatically. This capacity is reliant on the structure's adjustable Arduino light sensor detects fluctuations in the amount of direct sunlight the structure receives from any angle during the day. This can replace conventional curtains or blinds and is especially useful in places with consistent weather patterns.

Curtains and blinds are more expensive when taking into account the original cost of the goods as well as the time required for installation and upkeep. However, their effects on temperature regulation, sunlight exposure, and solar illumination are only slightly different depending on the weather. A further advantage is that, when sunlight is absorbed by the glass, it can eliminate the reflection of sunlight over the building, so converting the thermal pollution component into a regenerative renewable storable solar energy. Research conducted on the created prototype demonstrates that they are feasible and provide a cost-effective but effective option for automated energy-saving systems [36].

4.2 Liquid crystal polymers' electrical characteristics

A well-known approach for creating polymer-dispersed liquid crystal (PDLC) films is the polymerization-induced phase separation approach. It examines the impact of chemical modification on the dispersion of multi-walled carbon nanotubes (MWCNT) in liquid crystals and examines how these qualities affect the morphology, electro-optical properties, and conductivity of the PDLC films. Results demonstrated that the threshold voltage and the saturation voltage of PDLC films decreased with the increase of the injecting amount of MWCNT or chemically modified MWCNT because carbon nanotubes can enhance the electric field by decreasing the resistivity of the medium and improving the capacitance of the cells. It is evident that the PDLC films doped with the chemically modified MWCNT have a higher contrast ratio than the MWCNT [37].

In this research, the gel molecules were synthesized with liquid crystal polymerizable monomers and embedded into the polymer matrix. Three types of gel molecules, Dimethyl dibenzylidene sorbitol DMDBS, 4-di-p-methylbenzylidene-D-sorbitol MDBS, and dibenzylidene-D-sorbitol DBS were chosen as the ratios between the gel molecules and the polymer monomers to create dispersed liquid crystal thin films with low driving voltage. The transitions between transparency and scattering of the thin films were achieved, and their theoretical assessment was based on the concentration of the gels, the temperature at which the polymerization occurred, and the intensity of the polymerization light. These factors contributed to the procedure of the electro-optical properties of the composite system. The findings demonstrate that while gel doping significantly improves PDLC's electro-optical performance, contrast also reduces it at the same time [38].

Large surface areas, efficient micropores, and modifiable acidity/basicity of zeolites have made them the perfect substrates for immobilizing small-sized and highly dispersed metal species. The metal catalysts supported by zeolites have been extensively employed in various catalytic processes in recent times, owing to their exceptional shape-selectivity, superior temperature and hydrothermal durability, and outstanding action. This review provides an extensive overview of the state-of-the-art developments in the last five years for significant heterogeneous catalytic processes using zeolite-supported metal catalysts. This includes different reactions. Lastly, a few constraints that are present now and some perspectives for the future on the opportunity and challenge for this subject are highlighted. To fulfil the growing need for useful applications, it is anticipated that this review will stimulate more creative research on the synthesis and catalysis of zeolite-supported metal catalysts and further their future advances [39].

Recent advancements in the fields of unique micro-mesoporous materials produced by zeolite recrystallization are discussed in this review. Based on the degree of recrystallization, the materials are divided into three distinct groups: (i) micro-mesoporous zeolites (RZEO-2); (ii) covered mesoporous zeolites (RZEO-1); and (iii) mesoporous materials having zeolitic fragments in the walls (RZEO-3). The investigation of the synthetic techniques leading to various types of recrystallized materials is the main emphasis of the first section of the review. A thorough analysis of their structure, texture, and porosity with their acidity and diffusion properties is provided in the second section. The catalytic uses of recrystallized materials are covered in the final section. The benefits and drawbacks of alternative hierarchical zeolites and pure micro- and mesoporous molecule sieves are evaluated critically, and discussion is held regarding the prospects and viewpoints for the future [40].

Although exceedingly unusual, the finding of new nematic phases at thermodynamic equilibrium is of significant importance because of the extensive use of nematic liquid crystal-based display technologies. The amazing finding of polar order and gigantic ferroelectric polarization in a nematic fluid is one of the most significant findings in the 150-year history of liquid crystals and marks an important turning point in the study of soft materials. Following a brief overview of this nascent topic, we showcase the state-of-the-art in establishing the phase's molecular origins before delving into how molecular structure influences this phase's occurrence and suggesting future directions [41].

Because when the copolymer weave continues to shrink, the sampling A to G's respective barrier height and saturation voltage rise, so according to TETA concentration rose, and both the voltage output (Vth) and saturating voltage (Vsat) dropped. Or, to put it another way, a decrease in the polymer mesh immediately causes a boost in Vth and Vsat. It is widely known that decreasing the polymer mesh results in an increase in the mating surfaces here between polymeric and the liquid crystal, which raises the greater surface temperature and anchoring energy.

Figure 4.1.2 shows the corresponding contrast of specimens A to G also rises as polymer mesh content decreases, while the contrast decreases with increasing TETA content.

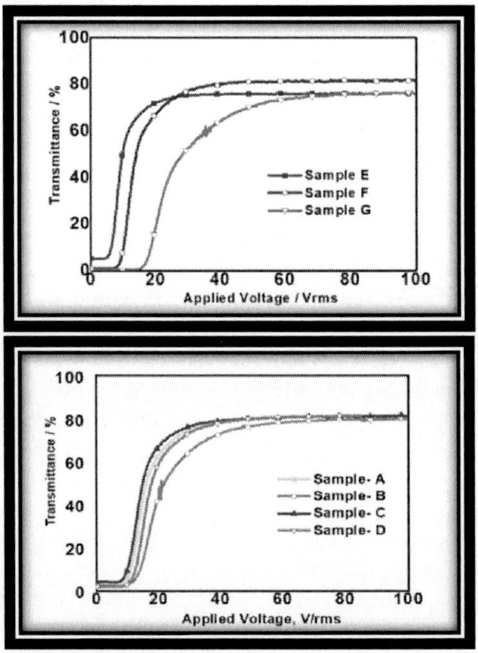

Fig 4.3 Applied voltage and transmittance

4.3 Influence of PDLC film response time

Figure 4.1.4 explains the relationship between the sampling A-G off-state situational awareness as well as the ratio of NGDE/EGDE and the application of a square wave voltage of 100V and 100Hz. As can be observed as the polymer mesh gets smaller, the on-state response time (t_{on}) of samples A through G increases, whereas the off-state response time (t_{off}) decreases.

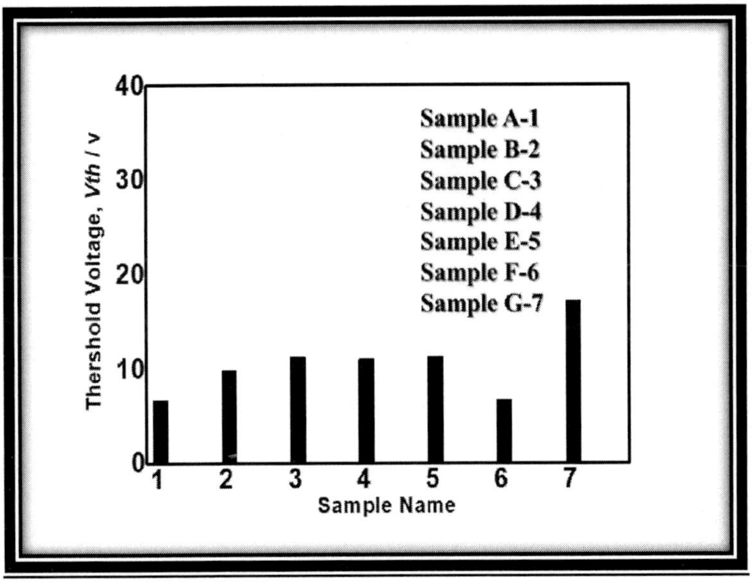

Fig 4.4 Threshold voltage (V_{th}) and saturation voltage (V_{sat})

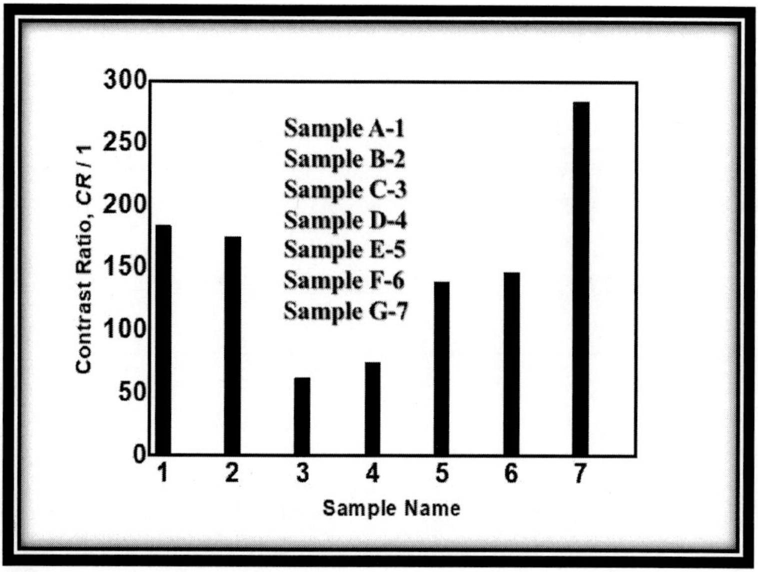

Fig 4.5 Contrast ratio (CR) of sample A-G

4.4 Off-state transmittance of PDLC materials

Interesting linear and nonlinear optical materials called liquid crystals are employed in a wide range of applications other than flat panel screens. With an electrical control, an optical control, or both, liquid crystalline materials can be employed as the foundation or cladding of switchable/reconfigurable waveguides. This work presents the materials and primary structures of devices of liquid crystals contained in various waveguide geometries utilizing a variety of substrate materials, including polydimethylsiloxane, silicon, soda lime, or borosilicate glass. About the optical anisotropy of liquid crystals, modelling of the behaviour of liquid crystal Nano metric molecule reorientation and its associated refractive index distribution under both low-frequency electric and intense optical fields is given. Several instances of waveguide-based integrated optical systems with liquid crystalline materials at their core for filtering and optical switching are being investigated. According to published results, one important aspect of photonic equipment based on liquid crystal light propagation is low-power control signals. These devices have competitive performance when compared to analogous integrated optic devices dependent on other materials for optical networks and optical sensing systems [42].

Acrylate-thiol is polymerized in two phases with wavelength selectivity to create polymer-dispersed liquid crystal (PDLC) films. This technique allowed for the successful preparation of the PDLC with a microsphere polymer shape, which significantly increased the adhesion and contrast ratio (CR) without degrading other performance parameters. By varying the Michael addition reaction's response time, the morphology could be changed from a porous to a microsphere-shaped structure. Furthermore, the free radical initiator's content may be adjusted to control the microsphere structure's size and homogeneity. This research should offer a useful technique for managing the acrylate-thiol polymers' microstructure throughout solution polymerization. It may also make it possible to create PDLCs with high CR and stickiness, which would be highly beneficial importance in light valves, switchable windows, and massive, flexible displays [43].

This work reports on the morphology, phase separation, instability, and dewetting of a polymer-dispersed liquid crystal (PDLC) thin film made of polystyrene (PS), 4-n-pentyl-4′-cyan biphenyl (5CB), and nematic liquid crystal. On topographically developed, flat silicon substrates, a thin layer of PS-5CB PDLC is spin-coated using a standard solvent (toluene). The study uses films with two different thicknesses, h = 35 nm and h ≈ 85 nm. The PDLC composition of RB (PS: 5CB) = 1:2 and 1:1 has been selected. Over time, at room temperature (T = 25°C), the thinner films (h ≈ 35 nm) on a flat substrate dewet and form isolated droplets for both RB. On the other hand, the h ≈ 85 nm thick films show a slow phase evolution of yet are stable at ambient temperature. These films only wet when thermally annealed at T ≈ 70 °C immediately following spin coating. This generates an isotropic collection of droplets with roughly similar sizes, each of which is composed of a 5CB ring around an inner polymer core. We demonstrate that casting the films onto a substrate with a topographic pattern allows alignment of both the instability patterns and the phase-separated LC domains. The 85nm thick film is continuous and experiences progressive phase evolution, producing an alternating domain of phase-separated LC (within the grooves) and PS domains aligning along the contours of the substrate patterns, in contrast to the 35nm thin film, that generates a spin dewetted array of aligned droplets along the substrate's grooves [44].

In this work, the Polymer Induced Phase Separation (PIPS) method is implemented to investigate the effects of cure temperature (Tc) and cure time (tc) on the morphology and a few electro-optical properties of flexible thermoset PDLC systems. We calculated the optical transmissions and switching voltages as functions of cure temperature and cure time, as well as the morphological characteristics, such as size, number density, and volume fraction of phase-separated liquid crystal droplets, under a constant thickness. We examined how the cure temperature and duration affected the electro-optical characteristics and shape of the thermoset PDLC films [45].

In this work, the UV-polymerization induced phase separation (PIPS) approach was used to generate a polymer-dispersed liquid crystal (PDLC) film with good electro-optical properties and a wide working temperature range. The curing temperature was optimized by employing an LC with a broad range of temperatures. The study discovered that the film exhibited improved electro-optical characteristics and a high contrast ratio of 51 and at 90°C when the polymerization temperature was at the prepolymer/LC mixture clearing point. Furthermore, the rise and decay times were found to be as fast as 241.5 ms and 1750 ms at −20°C, respectively. This study encourages the use of PDLC film in outdoor smart windows and offers additional recommendations for methods for the curing process [46].

The capability of polymer dispersed liquid crystal (PDLC) films to influence the reflectance of solar sails and hence control aircraft attitude has been the subject of much research. Unluckily, the majority of the polymeric matrix used in PDLC films is derived from polyacrylates, which are not able to withstand the extreme conditions found in space due to their lack of comprehensive qualities such as mechanical strength, chemical stability, thermal stability, and radiation resistance. In this research, polyimide-epoxy resin PDLC films were made by thermal polymerization utilizing two types of epoxy resin: monomer and Kapton polyimide film, which are conductive films to increase the radiation resistance of PDLC films. Studying the microscopic characteristics of PDLC films made from flexible and rigid chain epoxy resin monomers in varying ratios, the electro-optic properties [47].

Under either dry or wet conditions, the impact of interfacial interactions on the physical properties of polypropylene (PP)/natural zeolite composites was investigated. By surface treating the natural zeolite with a non-ionic surface modifier, such as 3-weight per cent polyethene glycol (PEG), and multiple types of silane coupling agents, such as 3-aminopropyltriethoxysilane (AMPTES), methyltriethoxysilane (MTES), and 3-mercaptopropyltrimethoxysilane (MPTMS), at four different concentrations (0.5–2 weight per cent), the interfacial interactions were improved to improve filler compatibility and mechanical properties of the composites. Polypropylene (PP) composites with a 2–6% weight percentage of zeolite were made using an extrusion method. It was discovered that the zeolite surface treatment affected the tensile properties of the composites, which were measured as a function of the filler loading and the coupling agent concentration. Silane treatment revealed notable enhancements in the mechanical characteristics of the composite. The PP composites with 1 weight per cent AMPTES-treated zeolite showed the greatest improvement in mechanical characteristics, as indicated by the dry and wet tensile test results. Pukanszky's semi-empirical equation was used to confirm the enhancement of the interfacial interaction. The prediction of the Pukanszky model and the experimental results showed good agreement. Studying the PP matrix with scanning electron microscopy also showed improved filler particles treated with silane dispersion [48].

In the zero- or micro-gravity environment of space, hydrophilic fin surfaces are necessary for condensing heat exchangers onboard manned spacecraft to enable soaking and wicking of condensate and ensure gravity-independent water separation. The coating needs to be biocidal to be able to stop germs from growing. Here, we demonstrate for the first time the exceptional hydrophilicity, biocidal qualities, and adherence of zeolite A and ZSM-5 coatings applied via in-situ crystallization to stainless steel and aluminium alloys. On the majority of the substrates examined, water contact angles were less than 5°. When the zeolite layer undergoes silver-ion exchange, it becomes extremely antimicrobial. Zeolite A's biocidal activity is regenerating due to frequent ion exchange. According to the adhesion test, every coating has a maximum rating of 5B. American Society for Testing and Materials, ASTM D-3359-02. Zeolite coatings have numerous benefits over the present sol-gel coatings, particularly their low-temperature crystallization process, corrosion resistance, and proven suitability for applications such as condensing heat exchangers onboard manned spacecraft.

The above-mentioned microstructure can be used to explain why, under the off-state condition, the light scattering intensity increases as TETA content increases. Since the liquid crystal droplets gradually get smaller as TETA content increases, the polymer network becomes denser under a microscope and the liquid crystals are dispersed in the matrix in the form of smaller droplets. As a result of this change, the PDLC film's light scattering is improved, and the off-state transmittance gradually decreases.

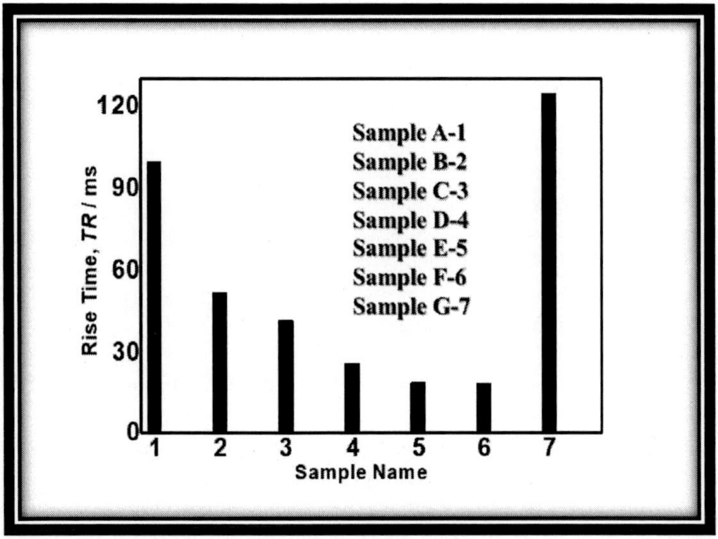

Fig 4.6 The time (T_{on}) time (T_{off}) of the sample A-G

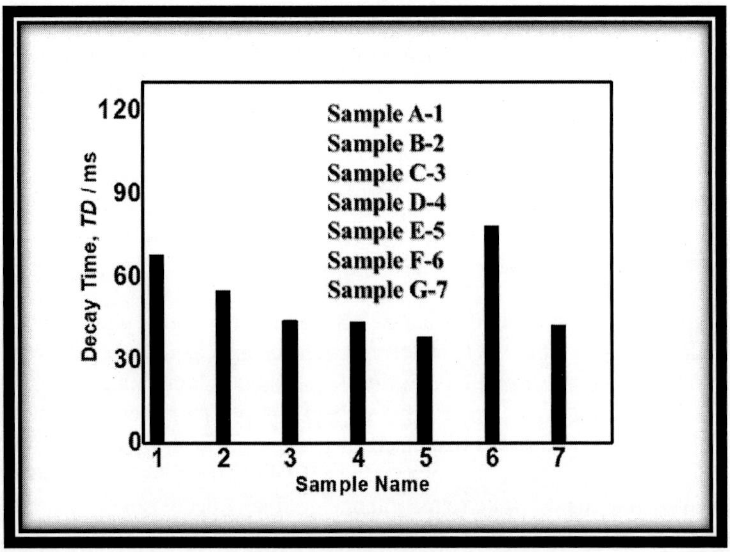

Fig 4.7 The time (T_{on}) time (T_{off}) of the sample A-G

Fig 4.1.7 shows that as the concentration of TETA increases intensity of scattering of light also increases and the rise and decay time is mentioned accordingly.

Conclusion

- Polymer dispersed liquid crystal (PDLC) films were prepared by polymerization induced separation (PIPS) method with nematic liquid crystal (LC) content as low as 20 wt% and the electro-optical properties were carefully investigated.

- The composition ratio, weight, and combined effects of curable epoxy monomers have significantly influenced the morphology and electro-optical properties of thermal-cured PDLC films.

- On the morphology Penta-functional monomer with two bulky methyl (-CH3) increases the crosslinking density on the polymer network of PDLC films regularly while increasing the catalyst concentration resulting in increasing the morphology of the thin flexible PET/ITO coated thin PDLC films.

- The electro-optical properties and morphologies of the PDLC films were strongly influenced by the chemical structure of curable monomers di functional ethylene glycol diglycidyl ether (EDGE) and penta functional Neopentyl glycol diglycidyl ether (NGDE).

- Moreover, the results presented in this study also suggest that it is possible to regulate the LC domain size and optimize the electro-optic performance by adjusting the composition ratio and structure of di and Penta functional curable epoxy monomers to obtain PDLC films possess good electro-optical properties.

References

Ailincai, D., & Marin, L. (2021). Eco-friendly PDLC composites based on chitosan and cholesteryl acetate. *Journal of Molecular Liquids*, *321*, 114466.

Sharma, S. C. (2010). A review of the electro-optical properties and their modification by radiation in polymer-dispersed liquid crystals and thin films containing CdSe/ZnS quantum dots. *Materials Science and Engineering: B*, *168*(1-3), 5-15.

Peng, S. C., Yu, J. W., & Lee, S. N. (1997). Effect of droplet size on the dielectric properties of PDLC films. *Journal of Polymer Science Part B: Polymer Physics*, *35*(9), 1373-1381.

d'Alessandro, A., & Asquini, R. (2021). Light propagation in confined nematic liquid crystals and device applications. *Applied Sciences*, *11*(18), 8713.

Katariya-Jain, A., & Deshmukh, R. R. (2022). Effects of dye doping on electro-optical, thermo-electro-optical and dielectric properties of polymer dispersed liquid crystal films. *Journal of Physics and Chemistry of Solids*, *160*, 110363.

Yu, M., Wang, T., Xu, J., Hu, W., Gao, Y., Zou, C., & Yang, H. (2022). Steric group-based polymer dispersed liquid crystal composite films with high contrast ratio, low driving voltage and small hysteresis. *Journal of Molecular Liquids*, *365*, 120152.

Islam, M. S., Chan, K. Y., Thien, G. S. H., Low, P. L., Lee, C. L., Wong, S. K., ... & Ng, Z. N. (2023). Performances of polymer-dispersed liquid crystal films for smart glass applications. *Polymers*, *15*(16), 3420.

Shaik, S., Gorantla, K., Mishra, S., & Kulkarni, K. S. (2020). Thermal and cost assessment of various polymer-dispersed liquid crystal film smart windows for energy efficient buildings. *Construction and Building Materials*, *263*, 120155.

Alghamdi, H., & Almawgani, A. H. M. (2019, May). Smart and efficient energy saving system using PDLC glass. In *2019 Smart City Symposium Prague (SCSP)* (pp. 1-5). IEEE.

Wu, Y., Cao, H., Duan, M., Li, E., Wang, H., Yang, Z., & He, W. (2018). Effects of a chemically modified multiwall carbon nanotubes on electro-optical properties of PDLC films. *Liquid Crystals*, *45*(7), 1023-1031.

Wang, L., Luo, S., Liu, S., He, Z., & Miao, Z. (2024). Mechanism of the effect of gel doping on the electro-optical properties of PDLC. *Liquid Crystals*, 1-15.

Sun, Q., Wang, N., & Yu, J. (2021). Advances in catalytic applications of zeolite-supported metal catalysts. *Advanced Materials*, *33*(51), 2104442.

Ivanova, I. I., & Knyazeva, E. E. (2013). Micro–mesoporous materials obtained by zeolite recrystallization: synthesis, characterization and catalytic applications. *Chemical Society Reviews*, *42*(9), 3671-3688.

Mandle, R. J. (2022). A new order of liquids: polar order in nematic liquid crystals. *Soft Matter*, *18*(27), 5014-5020.

Lu, H., Huang, P., Wu, T., Chen, C., Shi, J., Xu, M., & Zhu, J. (2022). PDLC with controllable microstructure using wavelength-selective two-stage polymerization. *Polymer*, *243*, 124641.

Dhara, P., & Mukherjee, R. (2021). Phase separation and de-wetting of polymer dispersed liquid crystal (PDLC) thin films on flat and patterned substrates. *Journal of Molecular Liquids*, *341*, 117360.

Hakemi, H. (2020). The effect of curing temperature and time on morphology and Electro-Optical properties of flexible thermoset polymer dispersed liquid crystal (PDLC) films. *Molecular Crystals and Liquid Crystals*, *703*(1), 1-12.

Xu, J., Li, H., Guo, L., Zhang, Z., Zou, C., Wang, Q., & Gao, Y. (2022). Effect of curing temperature on the properties of electrically controlled dimming film with wide working temperature range. *Crystals*, *12*(11), 1551.

Zhao, Y., He, Z., Ban, Q., Li, K., Ding, S., Tian, S., & Miao, Z. (2021). Study on the electro-optical properties of epoxy resin based polymer dispersed liquid crystal films. *Molecular Crystals and Liquid Crystals*, *722*(1), 1-7.

Metın, D., Tihminlioğlu, F., Balköse, D., & Ülkü, S. (2004). The effect of interfacial interactions on the mechanical properties of polypropylene/natural zeolite composites. *Composites Part A: Applied Science and Manufacturing*, *35*(1), 23-32.

McDonnell, A. M., Beving, D., Wang, A., Chen, W., & Yan, Y. (2005). Hydrophilic and antimicrobial zeolite coatings for gravity-independent water separation. *Advanced Functional Materials*, *15*(2), 336-340.

CHAPTER 5

5.1 Introduction to liquid crystal

5.1.1 A brief history of the development of liquid crystals

In 1888 Austrian botanist Friedel Reiter discovered a phenomenon: he was measuring Cholesterol benzoate at the melting point(C34H50O2) it was found that when heated to 145 0C, the crystals began to melt, first turning into a cloudy liquid and continuing to heat to to179 0C before gradually turning into a transparent liquid. This is accompanied by a change in colour [1]. The following year, German scientist Otto Lehmann studied the phenomenon carefully and used a polarizing microscope to find that the turbid liquid had an optical anisotropy like crystals. Indicates that this is a new phase between anisotropic crystals and anisotropic liquids of matter, and names its Liquid Crystals.

The discovery of the new phase state of the substance attracted many other researchers, but before 1965, the scope of research was only the synthesis of liquid crystal compounds and their phase-changing behaviour. Daniel Verlander of Germany proposed the first rule of thumb in 1908: the molecular structure of compounds that produce liquid crystal states should be as rod-like as possible, and the synthesis of numerous liquid crystal compounds, which reduces the blindness of synthesis, and he began to study the aggregation state structure of liquid crystals. British chemist G.W Gray synthesized more liquid crystal compounds in his team [2].

These provide a theoretical basis and experimental methods for the research and application of liquid crystal materials. In the early 1930s, W Kast, G. Friedel, C.W. Osteen et al. [3] created the liquid crystal continuant theory, studied the influence of the outer field (electric and magnetic field) on the liquid crystal, and determined the conductivity of the liquid crystal. Fredericks et al. discovered that the action of electromagnetic fields caused the deformation of the column phase and its threshold, and named it Fredericks. The Fredericks transformation is the basis for the application of liquid crystal displays (LCDs) as seen today. During the same period.

For the first time, Kast et al. have measured the dielectric constant of the liquid crystal. Before the 1960s, that is, more than 80 years after the discovery of liquid crystal, because there was no practical value, liquid crystal research was basically in the basic research stage [4]. It was not until 1971 that the racialization of liquid crystals was a breakthrough. This year Schact and M Helfrich discovered the matching of the distorted electro-optical effect of liquid crystals with integrated circuits and invented the distortion-to-column liquid crystal display (TN-LCD, Twisted Nematic-Liquid Crystal Display) and the dynamic optical scattering mode liquid crystal display (DSM-LCD, Dynamic Scattering Model-Liquid Crystal Display) [5]. The emerging liquid crystal industry at that time laid the foundation, for the display industry has brought a huge impact, but also greatly promoted the basic theory of liquid crystal research so that liquid crystal has been widely used. DSM-LCD is an application current-voltage effect that consumes more power relatively, while TN-LCD is a voltage-sensing drive with very little power consumption, so TN-LCD quickly took over the electronics market field [6].

Market demand has promoted the development of liquid crystal display research. In 1985T Schiffer and Nearing and G. Waters and others invented ultra-distorted column-phase liquid crystal displays (STN-LCD, Super Twisted Nematic-Liquid Crystal Display) which quickly captured the market and became mainstream products. In 1974, Meyer foresaw the theory of spontaneous polarization of the helix's hand-near crystal. In 1975 he synthesized DOBAMB can experimentally confirm spontaneous polarization, resulting in the discovery of ferroelectric liquid crystals [7].

In 1980, Clark and Lager well confirmed that the electro-optical properties of ferroelectric liquid crystals (FLDDs) were the domain of their research. The most important features are high-speed response and memory. In 1988, the Fukuda research team at Tokyo University of Technology in Japan discovered the first anti-iron-electric liquid crystal molecule MHPOB, which has an electronically controlled three-steady state. In short, in the past three decades, the research of liquid crystal application has been developed by leaps and forwards, and the basic research of liquid crystals has also achieved fruitful results.

The field of study spans many disciplines, such as physics, chemistry, electronics, biology, etc. The so-called liquid crystal refers to some substances under heating or cooling conditions that are not directly from the three-dimensional anisotropic crystal to the completely anisotropic liquid, but through a series of intermediate states, this intermediate state is called liquid crystal. The liquid crystal state of the substance has both liquid fluidity and crystal anisotropy [8]. When the liquid crystal phase changes, there are strictly determined radon and entropy changes, so the liquid crystal state is a thermodynamically stable intermediate state. The temperature T1 when anisotropic crystals are converted from anisotropic crystals to liquid crystals, as shown in Figure 5-1-1, is called Melting Point, and temperature T2 when it is converted from liquid crystals to anisotropic liquids is called Clear Point. The temperature interval between T1 and T2 is the liquid crystal phase range. According to the formation conditions and composition, liquid crystals can be divided into thermal liquid crystals and dissolved liquid crystals. The thermal liquid crystal phase is caused by temperature changes and can only exist within a certain temperature range. A dissolved liquid crystal is a liquid crystal system consisting of compounds and solvents that meet certain structural requirements, consisting of two or more compounds [9].

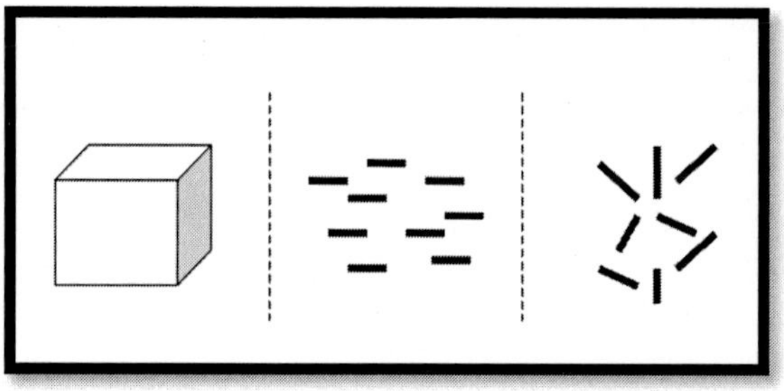

Figure 5-1 The phase change of liquid crystal matter

In the case of thermal liquid crystals alone, it was previously thought that to present the liquid crystal phase, the molecular structure of the compound must be rod-shaped, with several aromatic rings. However, with the further deepening of people's research on liquid crystals, the above laws have been broken, such as disc-shaped liquid crystals, single-benzene ring liquid crystals, banana-shaped liquid crystals, non-liquid crystals nuclear full-chain structures of liquid crystals and so on.

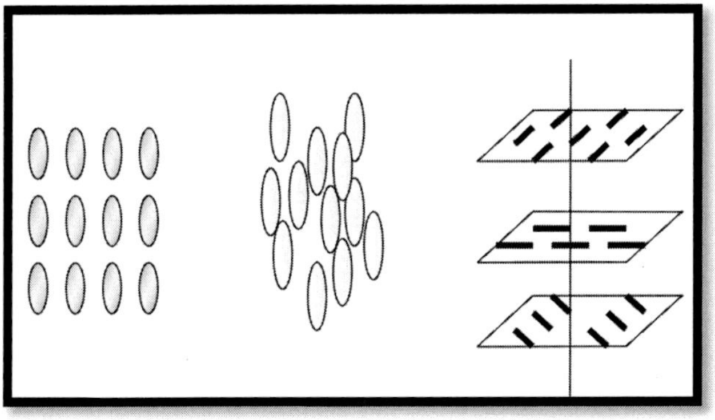

Figure 5-2 The molecular arrangement of the liquid crystal

Thermal liquid crystals can be divided into three types according to their molecular arrangement, as shown in Figures 5-1-2. Near-crystalline liquid crystal is composed of rod-like molecules arranged in layers, the long axis of molecules in the layer is parallel to each other, the direction of which can be perpendicular to the plane or tilted with the layer plane. The vertical phase liquid crystal is composed of rod-like molecules with a large diameter ratio, which is not arranged in layers and can only be kept parallel or close to parallel in the direction of the long axis of the molecule. Bile-phase liquid crystal is named after the bile steroid type, many bile-phase liquid crystals are a derivative of bile sterols. These liquid crystal molecules are arranged in layers, the molecules in the layers are parallel to each other, the molecular long axis is parallel to the layer plane, the direction of the molecular long axis of different layers is slightly changed, and the normal direction of the layers is arranged into a spiral structure.

5.1.2 Polarization chip characteristics of liquid crystals

Light is known to be an electromagnetic wave that refers to the direction of the electric field as the polarization direction of light. We can use polarization tablets to select polarized light in a particular direction. Polarized light can pass through or partially pass through or fail when passing through the polarization plate, depending on the direction of the polarized light and the direction of polarization of the polarizing chip (Figure 5-1-2). The qualitatively illustrates this function of the polarization chip [10].

When light passes through matter, the refractive index is related to the direction of light polarization. In most column-type liquid crystals, the refractive index is larger when the direction of light polarization is the same as that of the long axis of the molecule, and the refractive index is smaller when the direction of light polarization is perpendicular to the direction of the long axis of the molecule [11]. The direction of polarization of a polarized light sometimes changes after passing through a liquid crystal. Whether it will change or not depends on the arrangement of liquid crystals, as shown in Figures 5-1-3.

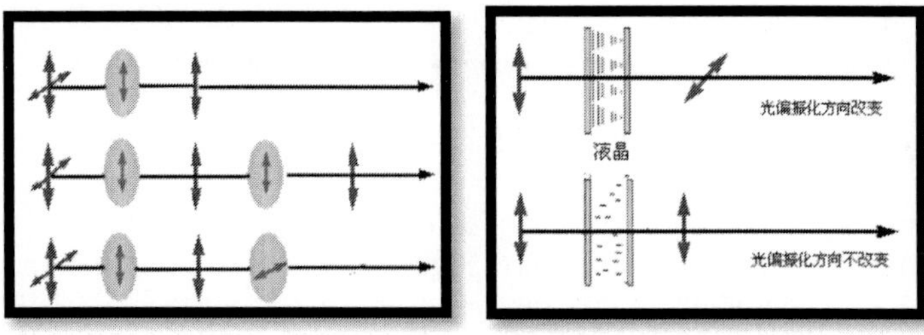

Figure 5-3 Polarization sheet application rationale

The molecule and the refractive index are smaller when the direction of light polarization is perpendicular to the direction of the long axis of the molecule. The direction of polarization of a polarized light sometimes changes after passing through a liquid crystal. Whether it will change or not depends on the arrangement of liquid crystals, as shown in Figures 5-1-3. Therefore, changing the arrangement of the liquid crystal can change the polarization through light. If you add another polarization chip to the back of the liquid crystal, the light force passing through will change. The polarization principle of the liquid crystal (2) is the dual refractive index of the liquid crystal [12]. The phenomenon of a beam of light splitting into two beams of light after it is shot into the liquid crystal is the double refraction of the liquid crystal, as shown in Figures 5-1-4. This phenomenon indicates that the dielectric constant and refractive index are different in different directions in the liquid crystal. Most liquid crystals have only one optical axis direction, the direction of light in the liquid crystal along the optical axis, there is no double refraction phenomenon, general liquid crystal optical axis along the molecular long axis direction, the photo shaft of the bile phase liquid crystal perpendicular to the layer.

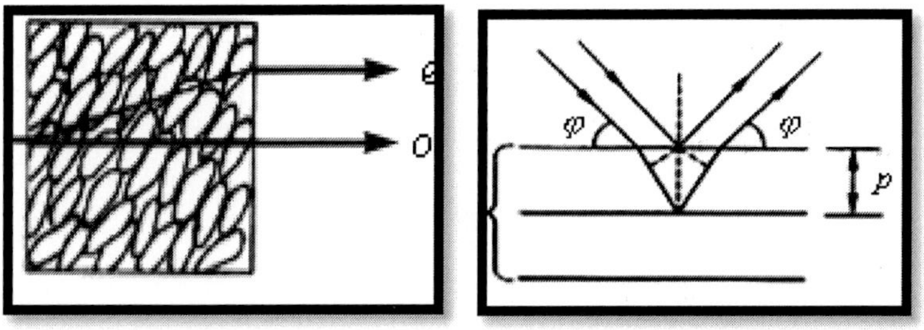

Figure 5-4 Double fold of liquid crystal

5.1.3 Application of liquid crystal

The phase liquid crystal presents a beautiful colour under white light, which is the result of its choice to reflect certain wavelengths of light. What wavelength of light is reflected depends on the type of liquid crystal and its temperature, as well as the incident angle of the light. This can explain this phenomenon of liquid crystals in the Prague formula of crystal diffraction, the φ of the crystal diffraction, and the 2npsin [13-19]. The medium is the wavelength of the reflected light map representing the pitch of the bile phase liquid crystalline is the average refractive index, and the φ is the angle between the incident light and the liquid crystal surface (as shown in Figures 1-6). From this, the light of different colours can be observed at different angles. When the temperature changes, the pitch of the bile phase liquid crystal changes sharply, and so does the colour of the reflected light. This feature of bile-phase liquid crystals is widely used in thermometers and in various display devices that measure temperature changes.

Liquid crystal is now widely used in various fields, such as physics, chemistry, electronics, and life sciences. Application of liquid crystal in physics: Because liquid crystal shows anisotropy in optical properties, some also have light activity, which can change the polarization direction of light. Physicists have pointed out that liquid crystal devices have important application prospects in the following areas [20].

5.1.4 Optical guide liquid crystal light

The ZnS layer exhibits lower resistance than that of the liquid crystal. When it is exposed to strong ultraviolet light, the resistance of the light guide layer decreases, enabling it to become lower than that of the liquid crystal material. This process facilitates the operation of the liquid crystal display

[21].To improve the sensitivity of the display, selenium is usually evaporated as a light guide layer, with an opaque, highly reflected light barrier layer placed between the light guide and the liquid crystal to prevent the light guide layer from being stimulated by visible light.

5.1.5 Optical modulator

Liquid crystal molecules are evenly arranged toward the column phase liquid crystal or bile phase liquid crystal, which are optical monoaxial substances, if these liquid crystals are applied electric or magnetic fields, then the orientation of the liquid crystal molecules will change, causing the optical axis rotation, and the liquid crystal box part of the application of electric field or magnetic field, the orientation of the liquid crystal molecules will become uneven, resulting in a partial refractive gradient. Using these properties of liquid crystals, biases and optical modulators can be manufactured [22].

Application of liquid crystal in electronics: Liquid has been widely used in electronic display devices, various sensors, optoelectronic components and other practical. LCD devices have controlled the market for other electronic devices to compete with it, it can be expected that the expected LCD TV will be popular, and LCD devices will also be more widely used. In addition, breakthroughs are expected in the following areas:

(1)Liquid crystal electronic light shutter: the liquid crystal box is placed between a pair of polarization plates, one side of the transmission to the liquid crystal box to apply voltage, polarization plates, one side of the transmission to the liquid crystal box to apply voltage, at this time, by selecting the liquid crystal molecular orientation pattern and polarization light plate combination, can be obtained from the dark field to transmission state of the transition.

(2) Micro-temperature sensor: in the liquid crystal box where horizontal orientation treatment is carried out, the arranged tissue formed by the mixture of column phase liquid crystal and bile phase liquid crystal is a spiral structure in which the molecular axis is parallel to the substrate and twisted in turn, and its pitch varies significantly with temperature, which is used to create micro-temperature sensors [23].

(3) Non-destructive detection: bile phase liquid crystal is very sensitive to temperature, a small temperature difference will cause its pitch to change, causing the wavelength of reflected light to change, and reflected light

colour to be different. In the aviation industry, to reduce the weight of aircraft, many components of the internal structure defects, the use of liquid crystal material detection, is very convenient [24].

Application of liquid crystal in chemistry: The application of liquid crystal in the chemical industry is very extensive, it can be used as an orderly solvent to promote the organic chemical directional reaction, stereoisomer selection, spin light material enrichment and separation, as chromatography fixing liquid can improve chromatographic selectivity and separation choice. These are important research topics in the chemical industry. The chemical application of liquid crystals is expected to breakthrough in the following areas:

(1) Trace gas monitoring badge: The use of liquid crystal physical and chemical effect, liquid crystal if adsorption of a small amount of chemical poison, its colour will change. Using this property, not only personal wear badges but also can be made into liquid crystals hanging on the wall, once the toxic gas escapes, the liquid crystal immediately discolouration, a warning signal. The chemical application of liquid crystals is expected to breakthrough in the following areas:

(2) Trace gas monitoring badge: The use of liquid crystal physical and chemical effects, liquid crystal if adsorption of a small amount of chemical poison, its colour will change. Using this property, not only personal wear badges but also can be made into liquid crystals hanging on the wall, once the toxic gas escapes, the liquid crystal immediately discolouration, a warning signal.

5.1.6 Application of liquid crystal in life science

The research work on bio-liquid crystals has achieved fruitful results, and the structure and principle of liquid crystals have been used to explain a wide range of life phenomena, including human beings, animals, plants, and microorganisms. At present, scholars from all over the world are making extensive and dynamic explorations of bio-liquid crystals, and their research work is developing along with both theoretical and practical aspects.

(3) Night vision bionics

Using advanced electronic technology to receive the faint light reflected by objects at night, and then use various ways to enlarge, and finally on the fluorescent screen to obtain a higher brightness image. At present, all kinds of night vision instruments, are only the light to carry out a simple amplification effect, and night action of the retina, as well as the image of the role of information processing, which is also incomparable to night vision. In this sense, the mystery of the increased sensitivity of the night animal's eyes is worth further exploration.

(4) The liquid crystal structure of the complex

The structure of the cornea crystal of the double eye, like a deflector, the micro-tube structure of the photoreceptor, and like a polarizer, through these two sets of "light detection devices", the insect compound eye can check the polarization measured. Of course, there are many scholars in this field with great interest in vigorously carrying out the research work on insect re-eye "navigation" mechanism, they hope that the breakthrough in this regard will design a more accurate navigation and positioning installation, to provide new ideas. Liquid crystal is used in industry as liquid crystal display technology, it can be called the iconic application of the liquid crystal. Liquid crystal as an information display material, compared to other materials, there are many incomparable advantages: first, low power consumption [25].

The power consumption of the liquid crystal display is many times less than that of the luminous material; Third, can be displayed in a bright environment, but the brighter the exterior light, the clearer the display characters, images, and finally, LCD device size flexible, novel style, to achieve device flat, curved, easy to use, low cost, in addition to it has a wide range of use, as well as storage capacity, choice and other advantages. The first research on liquid crystal display applications was the Williams and the A.P. of the former Soviet Union Kapustin's research on the formation of stripes (Williams) and discovery of light scattering in the action of the column-phased liquid crystal electric field has since evolved into the discovery of dynamic scattering mode (dynamic scattering mode: DSM) and the application of liquid crystal display LCD (Liquid Crystal Display).

The application of DS mode in the electronic display has had a great impact on the electronic industry and has produced colour television sets, electroluminescent, colour video recorders and so on [26]. Since 1970, watches and calculators in DS mode have been developed in the United States and Japan. In addition, healer et al.'s guest mode, Xerox J. Adams, G.Variable double refraction (TB), also known as ECB electronically controlled double fold, for Assoiling (1971). And, most importantly, J.L.

Ferguson M. Schactand W Helfrich's twisted-column Twisted Nematic: TN-LCD.

In the early 1970s, scientists developed DSM-LCD watches and calculators. However, because of the voltage effect of DSM-LCD applications, which consume more power, people are turning to TN-LCD products. S. Kobayashi and others made TN-LCD defect-free colour polarization chip display and made it rapidly industrialized, which is widely used in calculators, watches, electrical displays and so on. In 1973, Professor Gray (University of Hale, UK) discovered a stable liquid crystal material (biphenyls) [27-31].

In 1976, for the first time in the world, SHARP applied it to the display of the Calculator (EL-8025), which is now the foundation of LCD materials. LCD in television and computer research directions increases the amount of information displayed, increases the display area, and improves readability, full colour and so on. Brody proposed active-matrix modes to improve the readability of LCDs; T.Schiffer et al. invented super-distortion (STN) LCDs (1983-1985); N. Clark et al.(1980), K. Yoshino (1983)invented ferroelectric Liquid LCDs to increase the amount of information displayed. Since then, the GH LCD has also been studied, such as GH iron-electric liquid crystal display with dual stability and high-speed responsiveness characteristics [31].

At present, liquid display technology has been widely used in our surrounding clocks, calculators, televisions, word processors, personal computers, telephones, and other fields. In the information age and the deepening of internationalization, information equipment has been more and more widely used in the industrial sector and ordinary households, so the demand for information display equipment has also increased, and strongly demands its diversification, personalization and advancement. In this way, the demand for liquid crystal displays is increasing, and the requirements of liquid crystal displays are getting higher and higher, and to light, thin, low-consumption point, large-scale integration direction.

Japan is among the world's leading researchers in the development and application of liquid crystal technology. Even with overcapacity, equipment investment in Japan's chemical industry increased slightly in 2003, driven by high-value-added products such as materials for thin film displays. Thus, the liquid crystal industry occupies an important position in the Japanese chemical industry. The LCD industry in Korea and some countries in Europe and the United States has also followed Japan's, and some aspects have even surpassed Japan's, such as South Korea's TFT production technology has been very mature. The development status of the liquid crystal industry at home and abroad [32].

In recent years, the development of liquid crystal materials has been very rapid, almost to the point of rapid change. Liquid crystal polymer is a class of high-performance engineering plastics developed in the ages, although it started relatively late, but developed very rapidly. So far, the world has several companies that have achieved industrialization, annual output has exceeded 10,000 tons. The main producers are the United States, followed by Japan and Western Europe. At present, industrialization or the process of industrial development is mainly limited to the Eco critical base group located in the main chain, phase-to-column type of thermal type [33].

In recent years, the University of the United States has studied the latest liquid crystal polymer composite film materials developed by others. The continuous phase-in is the liquid crystal. For optical valves and display devices to show excellent electro-optical performance: and good transparency through the state, the liquid crystal and high polymer refractive index matching requirements are not as high, so in the transmission state not only the transmission rate is large and fog-free.

The perspective is wider and is often described as "full view". Its low drive voltage, short response time, fast transformation, small lag effect, and good grayscale, can achieve multiple drives. Reverse has good electro-optical performance. The electro-optical performance and related application research are in-depth. The research of liquid crystal display technology in China started at the same time as the world, but the realization of liquid crystal display industrialization is after a few years. Our country, - liquid crystal materials have lagged foreign countries, especially Japan, probably.

There is a five-year gap. At present, there are many LCD production lines, LCD device manufacturers are more, but only limited to products, high-grade product lines can only be imported from abroad, and the number is small. The development of China's liquid crystal industry. The main problem facing the exhibition is: the product technology is low grade, most of them can only produce products, but at the high end of the and so on is weak. China's liquid crystal display technology research started in the year and started at the same time as the world. But the real formation of the liquid crystal display industry is after the years, so far, China's liquid crystal industry has generally experienced four stages of development [34].

Phase 1: In the first phase: The Electronics Department, the Academy of Sciences and the Shanghai Electronic Tube Factory have introduced the production line of plus-inch substrate glass, mainly for watches, calculators, and some instruments liquid crystal products. At present, these production lines have no production value, or discontinued or have long been modified.

In the second stage: of the year, Changsha factory, Shenzhen Tiana, Shenhua, Shijiazhuang electro-optoelectronics, Fujian Anhui and other lead-inch production lines, most of the plant is still in production. Phase III: Beginning, introduction, inch production lines, such as Tiana II, Changhua, Xinyi I, He yuan Seiko, Shenhua II, Jingle, etc., these lines yield large, more advanced equipment, high yield, better efficiency, are currently the main production line.

Phase 4: Years later, China began to introduce or inch production lines, of which investment of 10,000 U.S. dollars from the United States to introduce equipment, only suitable to produce point screens and small and medium-sized screens. Investment of more than 10,000 U.S. dollars to introduce Japanese complete sets of equipment manufacturers have Shenzhen Tiana, Hebei Yaya, and Yawned letter sharp. These production lines have a high degree of automation, production beat fast (-piece), and plant purification conditions, with the production of high-grade, large-inch conditions, but in addition to Wuxi Sharp, due to technology, it is difficult to mass production of high-grade products.

After years of efforts, China's industrial use of basic materials from scratch has gradually formed a considerable scale of industry, including liquid crystal materials and the conductive glass localization rate is relatively high, the supply of other materials will gradually turn to the domestic.

(1) Liquid crystal materials If Tsinghua University in Shijiazhuang set up a joint venture LCD materials factory, the production of liquid crystal count, has been years. Shijiazhuang's strength LCD company sales of tons of liquid crystal, sales near million yuan. At present, in China's production of liquid crystal materials in addition to general materials, there are materials to meet the special requirements of wide temperature and low threshold and sub-type materials. The liquid crystal material domestic rate can reach the above, but many foreign liquid crystal inflows have created pressure to reduce prices. (2) Conductive Glass Shenzhen South Asia is the first company from abroad to introduce a conductive glass production line. But at that time several productions were very small, and the technology was backward. Since the beginning of Shenzhen Labia, Nano, Weigand and other

companies have introduced foreign advanced vertical continuous coating production lines, after digestion, absorption, and transformation, so that the output and quality of conductive glass have been significantly improved. China has a total conductive glass production line, which is introduced with the domestic production line in half, with a total production capacity of about 10,000 pieces per month ("standard calculation)."

The main challenge facing the development of China's liquid crystal industry is the low grade of product technology. Most manufacturers are only capable of producing basic products, while they struggle to compete in the higher-end segments. This limitation hinders the industry's overall growth and innovation, making it difficult to keep pace with global advancements in liquid crystal technology. As a result, there's an urgent need for investment in R&D and talent development to enhance product quality and technological capabilities. They are:

(1) Product technology grade is low, our country's output accounted for the world market above, but sales are not enough. The main reason is that most of them can only produce products. Not only a small number but also can only produce low- and medium-grade products, the current market demand for mobile phones, Chinese English translation machines and other high-grade products cannot be mass production, basically rely on imports. The market-occupied products have not yet been produced.

(2) Most enterprises produce too small, single products, and poor equipment, in the current market excessive competition conditions, and many enterprises can only maintain simple reproduction for a long time, difficult to achieve development.

(3) Lack of macro-control, repeated introduction of too much, resulting in the dispersion of capital, technical forces, who are not good, not only caused a large amount of waste of capital, not to build some strong, competitive first-class enterprises.

The LCD industry has high-tech, low-cost, high-efficiency characteristics. If it can be reasonably developed, it can play a big role in the country's economic development. On the one hand, the liquid crystal industry has high technical content, and its production costs are relatively low, as a high-precision industry, can obtain very high economic benefits; To sum up, we should increase the research efforts in liquid crystal and liquid crystal-related content, and promote the development of China's liquid crystal industry.

The development direction of liquid crystal display is mainly reflected in the following three aspects:

(1) Increase the viewing angle of the liquid crystal display. To solve this problem, the current methods are membrane compensation, multi-domain vertical arrangement and plane drive.

(2) Reduce the drive voltage, speed up response time and improve contrast. On the one hand, the existing liquid crystal materials can be improved, on the other hand, new liquid crystal materials can be developed, such as the response time of the material has been reduced to the next, can fully meet the requirements of dynamic display.

(3) Reduce costs and make LCDs popular. With the expansion of the LCD production scale, the improvement of production technology and the development of new liquid crystal materials, this problem will be solved.

References

Peter. J. Collins. **(2002).** Liquid crystal. *Shanghai Science and Technology Education Press*, : 35-50.

Zhou Qifeng, Wang Xinju. **(1994).** Liquid crystal polymer. *Science Press,* 165-88.Yukio Suzuki, translated by Kim Yu – Yu. **(2003)**. Color LCD display. *Science Press, :* 99-110. Kotshiko Kotsin,Qiao Shuang. **(2003)** Gao Yan translation Next-generation LCD display. *Science Press,*: 3-15.

Henry R M, Ramsey R A, Sharma S C. **(2004)** Effects of Crosslinking Agent Cure Temperature and UV Flux on the Electro-optical Properties of Polymer- Dispersed Liquid Crystal Cells. J Polym Sci Part B: *Polym Phys* 42: 404–410.

Song J H, Park L S, Eeon E T, Kim Y B, Tonmiki I. **(2003)** Functional Materials and Electro- optic Prosperities for Polymer Stabilized Ferroelectric Liquid Crytal. *Synthetic Metals,* 137:1411-1412.

Henry R M, Ramsey R A, Sharma S C. **(2004)** Effects of Crosslinking Agent, Cure Temperature, and UV Flux on the Electro-optical Properties of Polymer Dispersed Liquid Crystal Cells.J Polym Sci Part B: Polym Phys, 42: 404-410.

Yang Qishan, Sun Zhengmin. **(1993)** Liquid crystal display manufacturing process. *University of Electronic Science and Technology Press*, 4:1- 20.

Doane J W, Vaz N A, Zumers. **(1986)** Field controlled light scattering from nematic microdroplets, *Applied Physics Letters*, 48, 269-271.

Li Beiwei. **(2001)** China's liquid crystal industry development status quo and industrial development of technology choices, *technology economy* 160(4) : 26-29.

Neil B Cramer, Christopher N. **(2001)** Journal of Polymer Science: Part A: *Polymer Chemistry,* 29, 2311-2319.

Neil B, Cramer, Cramer, Sirish K Reddy. **(2004)** Journal of Polymer Science: Part A: *Polymer Chemistry*, 42, 5817-5826.

Kim S H, Chong P H. **(1998)** Effect of prepolymer structure on the electro-optic performance of polymer dispersed liquid crystals. *Polymer International*, 6(46):143-149.

Li Shanjun, Ji Caigui. **(1993)** the principle and application of polymer photo chemistry *Fudan University Press*, 1(55):105-109.

Zhu Junfei, Zhou Meihua. **(2003)** the production status of light primer Fine and specialized chemicals, 1(6): 13-19.

Gill G, Pojman N, Willis J A Whitehead J B. **(2003)** Polymer-dispersed liquid crystal materials fabricated with frontal polymerization. Journal of Polymer Science Part A: *Polymer Chemistry*, 41:204-212.

Abderrahim B, Smaïn. **(2003)** Temperature, and liquid crystal concentration effect on thermal conductivity of poly (styrene) dispersed 5CB liquid crystal. *Journal of Applied Polymer Science,* 89(2):481-486.

Joe B,Whitehead Jr,Nicole L,Gill. **(2001) Liquid** Crystals V, lam-Choon Khoo, Editor, *proceedings of SPIE Vol.4463*.188-197.

Kundu P P, Verma G, Raina K K. **(2003)** Influence of different polyesters and their molecular weight on the textural and electrooptical behavior of polymer- dispersed liquid crystals. *Journal of Applied Polymer Science*, 87(2):284-289.

Byung K K, Seon H K. **(1998)** Copolymer composition-dependent light transmission of polymer/liquid crystals composite films. *Journal of Polymer Science Part B: Polymer Physics,* 36(1):55-64.

Byung Kyu Kim. **(1998)** Journal *of Polymer Science: Part B: Polymer Physics*， Vol.36, 55-64.

Masashi K, Hideo F, Takeshi M, Hiroto S, Yosihide F, Masahiro K, Hiroshi K, Taichiro K, Fumio S. **(2004)** Fabrication of extended porous polymer films for aligning nematic liquid crystal. *Electronics and Communications in Japan (Part II: Electronics)*, 87(4):1-8.

Zhong Z Z, Schuele D E, Gordon W L, Adamic K J, Akins R B. **(1992)** Dielectric properties of a PMMA/E7 polymer-dispersed liquid. *Journal of Polymer Science Part B: Polymer Physics*, 12(30): 123-126.

Ren Hongwen, Huang X443-1449ixuan, Ling Zhihua, etc. **(1998)** The electric field orientation of liquid crystal molecules in polymer dispersion liquid crystal film should be liquid and display, 3: 33-35.

Park N H, Cho S A. **(2000)** Preparation of polymer-dispersed liquid crystal films containing a small amount of liquid crystalline polymer and their properties. *Journal of Applied Polymer Science*, 9(77): 3178~3188

Wu J, Wang C M. **(1997) Electro**-optical properties of aligned polymer dispersed liquid crystal films. *Physics Letters*, 21(7): 149-154.

Ren Hongwen, Ling Zhihua, Keita City Liang. **(1994)** A polymer dispersion liquid crystal (PDLC) membrane with high switching properties. *LCD Communication*, 2 （4):312-316.

Pogue R T, Natarajan L V, Siwecki S A, et al. **(2000)** Monomer functionality effects in the anisotropic phase separation of liquid crystal. *Polymer chemistry*, 41(1): 733-741.

Zheng Qixia. **(2000) Study** on the temperature characteristics of PDLC membrane light transmission rate. *LIQUID and display*， 15 （4): 255-258

Lin Yixin, Ren Hongwen. **(2000)** Distortion for high-contrast reflective displays, PDLC （5), 45-49.

Zhou Huilin, Pan Caiyuan. **(1996**) Study on the phase separation process of epoxy-based polymer dispersion liquid crystal material, 6, 287-290.

Gu Lingxuan, Zheng Jihong, Chen Gang, etc. **(2000)** Polymer Dispersion Liquid Crystal (PDLC) optical switch study. *Optical instruments*， 24 (3):28-33.

Yang Yuliang. **(1994)** Advances in liquid crystal /polymer composite display materials. *Journal of Polymer Chemistry*， 15 (12) ： 1868-1874.

Yu Tianshi, Lin Jiaping, Zhang Wenjie, etc. **(2003)** the phase behavior of polymer dispersion liquid crystal. *Polymer Materials Science and Engineering*, 11 (19):202-205.

CHAPTER 6

Abstract

It can be concluded that when the UV1 polymer content increases, themes of the PDLC membrane polymer network decrease, resulting in a gradual increase in its drive voltage and an increase in contrast as the content increases. Regarding response time, both T_{on} & T_{off} become shorter as the UV1 content increases. However, as the UV1 content increases, the drive voltage increases. Therefore, considering the comprehensive consideration, the matching of the moderate A2 sample effect is better.

The voltage-to-pass ratio graph from the PDLC film sample, the contrast change graph with temperature and the V_{10} and V_{90} with the prepared temperature change graph can be found. As can be seen from these diagrams, V_{90} of the PDLC film sample generally increases with the temperature of the film, the light scattering intensity of the off and open states decreases with the increase of the preparation temperature, the T_{on} changes rapidly with the rise of the preparation temperature, and the T_{off} decreases with the increase of the preparation temperature. The reason may be that when the temperature is low, in the mixing system near the bright spot, the polymer network of PDLC membrane samples is relatively small, so that the surface area of the interface between liquid crystal droplets and polymers increases, increasing the force between the liquid crystal and polymer network, so V_{10} increases, its contrast is also higher, the open response time is not much change, the fastest shut-down response time. With the increase of temperature, the mesh of the polymer network gradually increases, and some macromolecules of sulfur-containing alcohol space play a blocking role, thus affecting the threshold voltage and saturation voltage, the comprehensive overall data can be seen that the preparation of UV1/LC system PDLC effect at low temperature is better, the comprehensive parameter performance is higher, especially at 150^0C, the sample's electrical-optical performance is better.

6.1 Introduction to polymer dispersion liquid crystals

6.1.1 Liquid crystal and polymer dispersion species

Liquid crystal and polymer dispersion forms can be divided into liquid crystal dispersion in cured polymers (PDLCs and holographic-PDLCs) and polymerization networks dispersed in liquid crystals (network stabilized liquid crystals) (NSLCs). PDLCs (Figures 1-7) are tiny column-phase liquid crystal droplets in the polymer substrate evenly dispersed into liquid crystal droplets with size of a few microns or so. H-PDLCs (Figure 1-8) are polymerized in the case of two beams of light forming interferometry, in which the polymer induces the inner of a multi-layer liquid crystal to form a periodic structure consisting of column-phase liquid crystal droplets and multi-layer polymers, thus forming a three-dimensional hologram with a rotating light. PDLCs and H-PDLCs can control their light transmission and light scattering states by applying voltage externally Figure 6-1-1 [1-4].

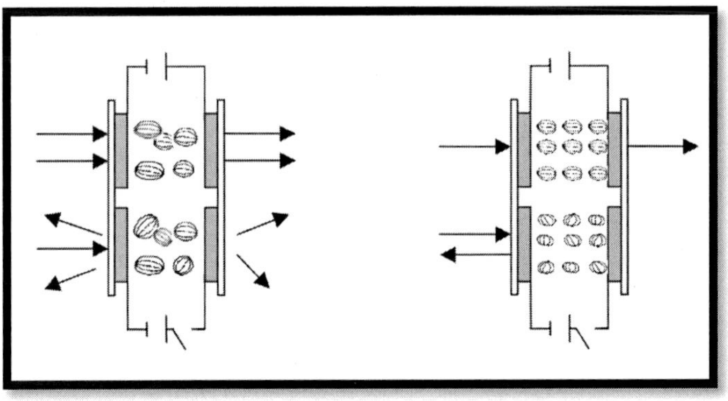

Figure 6-1 PDLCs illustration Figure1-8 H-PDLCs diagram

NSLCs and PDLCs, H-PDLCs differ in polymeric monomer ratios and liquid crystal selection in that NSLCs are distributed in the polymer network structure. As shown in Figure 6-1-1., the thick black line part of the figure represents the aggregation network, and the coke-shaped stripes represent the bile-phase liquid crystals.

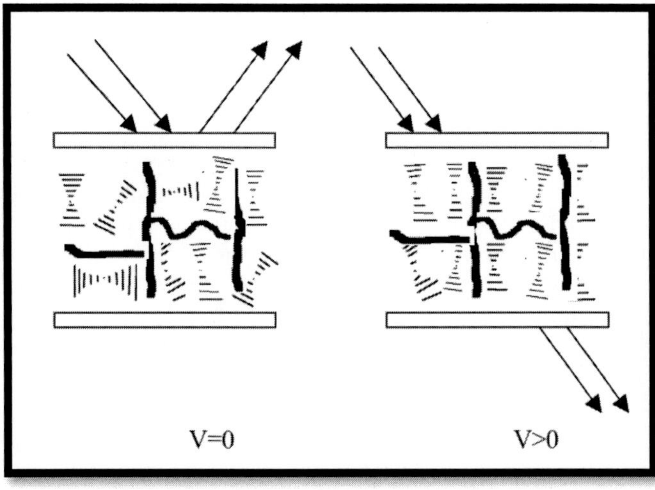

Figure 6-2 NSLCs polymer network dispersion

6.1.2 Introduction to UV aggregation

Light polymerization refers to any process by which a compound causes an increase in molecular weight due to the absorption of light, including further light crosslinking of pre-generated macromolecules, segment co-convergence and grip ion co-convergence caused by a certain light. So far, the optical polymerization found, in addition to photo-shrink polymerization (also known as local chemical polymerization), is, by its very nature, a chain reaction mechanism, that is, a chain-growth polymerization process caused by the active species (free radicals or ions) [5].

The difference between chemically induced free radical polymerization and ion-type polymerization is that the active species that trigger polymerization are produced differently. The active species of photoresist polymerization is produced by photochemical reaction, and the subsequent chain growth and chain termination processes are the same. Therefore, light polymerization only needs to absorb light energy during the chain-triggering phase.

Optical polymerization is characterized by the low activation energy required for polymerization reactions, so it can occur over a large temperature range, especially for low-temperature polymerization, which is much superior to chemically induced polymerization. In the laboratory, pure polymers that do not contain the residue of the triggering agent can be obtained by photo aggregation, which provides a very convenient means for further research [6].

Optical polymerization has provided a powerful means for the study of free radical polymerization dynamics and reaction processes. In addition, the photo regulation chain reaction is the process of absorbing a photon that causes many monomer molecules to converge into macromolecules. In this sense, optical aggregation is a kind of high quantum efficiency optical reaction and therefore has great practical value.

Optical polymerization requires that a component of the polymerization system must be able to absorb light energy at a certain wavelength range, and secondly requires that molecules that absorb light energy can further decompose or interact with other molecules to produce primary active species; The macromolecular chemical chain formed during the entire polymerization process should be able to withstand light radiation. Therefore, it is important to select the appropriate energy of light radiation so that it can produce the active species that trigger polymerization [7-9].

6.1.3 Principle of ultraviolet light polymerization

Most optical aggregations are done in free radical mechanisms. Some monomers are exposed to ultraviolet light and bond breaks to produce free radicals, called direct photolysis-inducing mechanisms. There are not many such monomers, such as alkyl vinyl ketones, bromine, etc. can absorb waves growing up to 300nm of light and directly dissolute into free radicals, which in turn trigger polymerization by free radicals. Light-induced gas phase polymerization with wavelengths from 200nm to 300nm has been applied to the formation of ultra-thin (20-100nm) polymer membranes on the substrate by positive and negative photoreceptors. When exposed to monomers such as styrene, methyl acrylates, and acrylonitrile at a wavelength of less than 300nm, they can only occur slowly [10]. This polymerization can be carried out in the gas phase, liquid phase, and in some cases solid phase. Direct light polymerization is not limited to the initiation of free radical chains, and other monomers that are generally easy to polymerize ionization, direct light polymerization may occur under vacuum irradiation.

6.1.4 The effect of aggregable monomer structures on aggregation

The physical properties of polymers associated with photo polymeric reactions are mainly polymeric molecular weight, chain flexibility, molecular motion transition temperature and aggregate state structure. It has been shown that excessive crosslink concentrations on polymer chains are not conducive to optical crosslinking. The reason may be that the crosslinking group along the polymer chain is overcrowded, and the initial rate of the crosslink reaction is too fast, but with the reaction, the crosslinking network formed by crosslinking is too dense, causing the residual official group to lose its degree of freedom and is not conducive to the further occurrence of crosslinking reaction, the crosslinking reaction rate may decrease rapidly. In addition, while changing the concentration of the reaction group, inevitably, the physical properties of polymers that have a greater impact than photosensitivity are changed. The molecular weight of polymers has always had a positive effect on the optical crosslinking reaction. In most cases, the photosensitive size of photosensitive polymers is still represented by a gel fraction [10]. The gel fraction is not consistent with the actual amount of light reaction, for some light crosslinking reactions, the gel fraction is related to the molecular weight of the polymer and its distribution, and the minimum amount of optical cross-linking that pre-reaches the insoluble system decreases with the increase of polymeric molecular weight. The sensitivity of the flexible chain segment is larger than that of the rigid chain segment, but the flexible chain polymer is not suitable as a photosensitive polymer material, because its mechanical properties and chemical stability may be affected after the reaction [11-14].

6.1.5 The type of light-inducing agent

Currently, only a few unsaturated monosomies absorb light in the 250nm-500nm range (they are the most readily available wavelength range in the laboratory). For most monomers, light-throwing agents or photosensitive agents must be used to trigger light polymerization. Even for the monomers listed above, that can be polymerized directly by light, light-inducing agents or photosensitive agents must still be used in most cases, considering the effective use of the energy of the light source and the degree to which the appropriate polymerization velocity angle is obtained.

The light-inducing agent is an indispensable component of UV-cured material, which plays a decisive role in the sensitivity of the optical curing system. Although Inmost obtained its first patent for ultraviolet (UV) curing in 1946, the concept of photo inductions began in 1967 after Bayer Argon industrialized UV curing technology. In 1970, Autocar-Cargill invented I2651, laying the foundation for the rapid development of light-inducing and establishing the company's leadership in the field of photo-inducing agent research and production [15].

Because the light-curing material is free of solvent volatilization when cured, it greatly reduces the pollution to the environment, so photocuring technology, as an environmentally friendly green technology, has been developed vigorously in recent years. According to statistics, in the past 10 years, the annual growth rate of foreign photo-inducing agents has been 81.1%, the annual growth rate of domestic UV curing products has reached 35%, and the number of light-inducing agents has also increased year-on-year. As an important part of photo-curable materials, the role of photo inductors is to absorb a certain wavelength of light energy to produce lively free radicals or cations, triggering or catalyzing the polymerization of the corresponding polymer monoliths.

As a light-inducing agent, it is usually necessary to meet the following conditions: (1) the quantum efficiency of the trigger, (2) the range of absorption spectrum matches the exposure to the light source, and (3) Good thermal stability, no dark reaction ;(4) and polymer monomers have better compatibility ;(5) light curing film without yellowing or discolouration;(6) safe and economical.

6.1.6 The principle of photoelectric effect

Under the action of the intricate electric field, the phenomenon of the change of the refractive index of the medium is the electro-optical effect, which can be expressed in the following form [16].

$$n\text{-}n_0=\alpha E_0{}^2\beta \text{ and } 2\cdots\cdots \quad [1]$$

Where n_0 and n are the refractive index of the medium before and after the action of the electric field, E_0 is the field strength of the additional electric field, and the α and β are constants. α E_0 is called a firster photo effect, also known as the Pock else effect, and βE_0^2 is called a secondary electro-optical effect, or Kerr effect. The first photo effect exists only in crystals that do not have a symmetrical centre, while the secondary electro-optical effect may exist in any form of crystal.

The main reason for the electro-optical effect: Under the action of the additional electric field, the original polarization form is destroyed, and the re-orientation occurs, thus producing stress and changing the physicality of the molecule. Materials with electro-optical functions tend to have both piezoelectric and ferroelectric properties.

Many factors affect the PDLC electro-optical effect, such as formulation, preparation conditions, droplet size and shape, self-organizing configuration, etc. In the PDLC membrane, liquid crystal is dispersed in the matrix in the form of spherical droplets, and when the liquid crystal molecules are oriented under the action of the electric field, they are restricted by the action of the liquid crystal/polymer interface, which not only acts on liquid crystal molecules near the surface of droplets but also interacts with each other Affecting the response of liquid crystal molecules inside the droplet to the electric field, the result is that the liquid crystal molecules limited by the droplets need a strong electric field to achieve the basic consistent orientation along the direction of the electric field when the electric field is removed, it can quickly revert to the initial state due to the effect of the interface. Due to the large liquid crystal/polymer interface in PDLC materials, PDLC operates at a higher voltage than conventional LCD devices and has an order of magnitude of 1-2 faster response times [17].

6.1.7 How PDLC works

PDLC is a liquid crystal with micron-scale small droplets scattered in the organic solid polymer substrate, each liquid crystal particle's optical axis is in the optimal orientation, while the light axis of different particles is random orientation, because of its refractive index anisotropy, so its effective refractive index does not match the refractive index of the substrate.

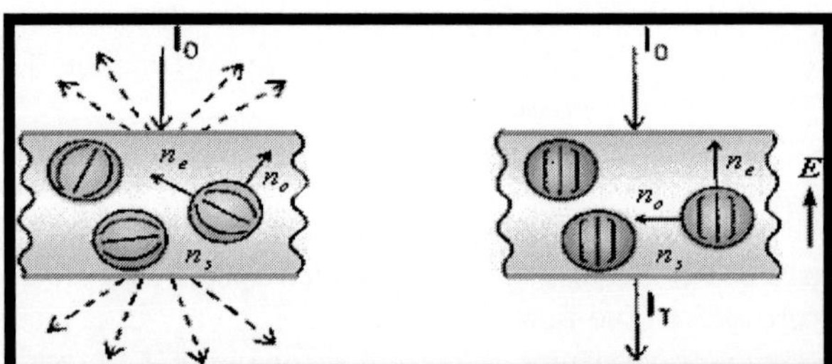

Figure 2-4 How it works is shown in

(a) Off state, the liquid crystal droplets are in a free-orientation state, the refractive index does not match the refractive index of the substrate, and the incident light can be strongly scattered into an opaque or translucent state. (b) Open state, when an electric field is applied, the particle orientation is parallel to the electric field and matches the refractive index of the substrate, and the light can pass through and appear transparent.

Under the action of an external electric field strong enough, the liquid crystal molecules in all particles are oriented along the electric field, when only the normal refractive index (n0) of the liquid crystal acts on light, and when its value is matched to the refractive index of the substrate, it appears transparent, as shown in Figure 6-1-4. Remove the external electric field, under the action of the elastic energy of the base wall, the particles resume the initial scattering1-10. Polymer Dispersion Liquid Crystals (PDLCs) are an emerging technology that offers significant advantages for display applications. One of the key benefits of PDLCs is that they do not require a polarization chip, simplifying the overall manufacturing process. This leads to lower production costs and greater ease of production compared to traditional liquid crystal displays that involve more complex processes such as distortion or hyper-distortion.

The optoelectronic performance parameters include threshold voltage, saturation voltage, contrast, response time, and so on. As a device for optical valves and displays, film response time or switching speed is critical in many applications, and the general switching speed is always limited by the time of liquid crystal droplets when the voltage is switched off. The response time of the control light valve is controlled by the following factors: the size and shape of the liquid crystal particles, the distortion constant and the value of the rotational stickiness coefficient, the interaction strength of the substrate and the liquid crystal molecule on the particles, etc. [18].

The size, shape and elastic constant of liquid crystal droplets, sticky coefficient and so on have an important influence on the switching time, threshold voltage, the threshold voltage, response time factors also have the dielectric anisotropy value of liquid crystal material, the strength of the external electric field. When the material is selected, the electro-optical characteristics of the liquid crystal particles are important in the size and shape, and the size and shape of the particles are obtained in the preparation process of thin film, so the study of this preparation process is also of great significance. Contrast By is a non-luminous display device, contrast is the most important parameter for measuring its display ability, defined as:

The penetration rate of the membrane in an open state and the rate of film penetration in a closed state. From this definition, obtaining a higher contrast can be achieved by selecting the appropriate substrate material and liquid crystal material, reducing the transmission rate of the closed-state film, another way to improve the contrast is to increase the thickness of the film, but this will make the driving voltage increase, and reduce the response speed. Therefore, the relationship between these and these should be handled comprehensively. In addition, to improve the contrast, de-chrome dyes can be mixed in the liquid crystal, so that the film has controlled absorption and reflection.

2.1.8 Drive voltage

The relationship between the pass rate and the voltage can be determined by applying a sine voltage action to both ends of the Indium tin oxide layer of the resulting film. Typically, the indicator that measures the size of the driving voltage is the voltage value applied when the transmission rate is the maximum. Many factors affect the driving voltage, such as the thickness of the film, the size of the liquid crystal particles, and the type of liquid crystal and substrate. The drive voltage is directly related to the energy consumption, cost, and practicality of the product, so it is a relatively important performance indicator.

6.1.9 Response time

Response time mainly includes open response time and off response time, open state response time is to apply a certain voltage value on the film when the pass-through rate reaches the maximum time required, response time is directly related to the membrane opening speed, which is particularly important in the display application.

The off-state response time is the time required to power off the film that applies a certain voltage value when the pass rate drops to 10% of the maximum value. Response time is affected and restricted by many factors, such as film thickness, type of liquid crystal and its content. Existing studies have shown that reducing the film thickness can make the response time shorter, and some fluorine-containing liquid crystals are used, because fluorine atoms have a higher electro nationality, so that the structure of fluorine-containing liquid crystals still has a certain dipole moment, viscosity is small, so the response speed becomes faster.

6.1.10 Preparation of polymer dispersed liquid crystal (PDLC) films

The Known preparation methods can be summed up as temperature Induction Phase: TIPS, Solvent inductive Phase Separation: SIPS, microcapsule method (Microcap's process: MP), and Polymerization Induced Phase Separation: PIPS. TIPS method is mainly the use of thermoplastic resins. If the melting point of the polymer is below its decomposition temperature, the PDLC film can be obtained using thermal phase separation. In this method, the polymer and liquid crystal are heated to the polymer.

Above the melting point, the liquid crystal molecules and molten polymer are mixed evenly, and then at a certain cooling speed to cool, liquid crystal with the polymer cooling hardened to form micro drops, the polymer and liquid crystal can be separated. In general, the size of liquid crystal droplets increases with the increase of the glassing temperature of the polymer. However, this method is more demanding on the membrane material, the resulting membrane is more sensitive to temperature, and less application of the MP method is the polymer and liquid crystal under the action of the carrier mixed, through strong stirring, the liquid crystal in the polymer species to form microcapsules. The size of liquid crystal droplets is determined by the stirring speed, generally with water as the carrier, but water removal is difficult, and time-consuming, thus affecting yield, and the presence of water will cause the electrochemical degradation of liquid crystal, its preparation process is complex, post-treatment difficult, less application [19].

SIPS method is applied to resins that cannot be processed with heat (melting point above decomposition temperature). That is, when the solvent is heated and volatile, the liquid crystal dissolved in the polymer solution is separated from the polymer to form particles in the polymer substrate, particle size is related to the rate of solvent volatilization, but the process of this method is much more complex than TIPS, and requires polar solvent recovery difficulties, but also cause environmental pollution.

PIPS method, the small molecular liquid crystal and pre-polymer are mixed in proportion to the desired ratio into a home solution, according to the polymerization effect to separate its phase, the polymerization method can be divided into epoxy thermoset method and ultraviolet light solid method two categories. At present, the method is more widely used, has more in-depth research and is a more mature technology, so we use the ultraviolet polymerization phase separation method in the PIPS method.

6.1.11 Application of polymer dispersion liquid crystal (PDLC)

The light valve made with PDLC does not need a polarization plate in its work, can reduce the absorption loss of light due to polarization, the angle of view is large, the display brightness is good, and in its production, the process does not need to surface treatment of the substrate, applied to the TFT display device production greatly improve the yield. The response speed is 1-2 orders of magnitude higher than that of TN devices, and grayscale is easy to control, greatly improving the display quality of the image and creating high-intensity, high-resolution projectors.

Because it is a solid-state membrane the liquid crystal leakage and film thickness control are very convenient and conducive to the production of large areas or bendable flexible displays. PDLC light valves can be used for electronic curtains and car roofs to control light incidents, keep secrets, and prevent cosmic rays. Since the shape of the particles can be "self-supported "by the polymer "self-supported" once the phase separation is completed during the processing process, the production process is simple, easy to control, low price, and can be manufactured in order of magnitude display devices and screens of square meters [16].

In addition to its use as a light valve in the display, PDLC can also be used as an optically nonlinear material for photosensitive protection, light conversion, and light calculation, as well as thermal and stress-sensitive components, so compared with TN components, PDLC has obvious advantages in terms of functional properties and processing performance. The studies have shown that ultra-thin and large areas of PDLC film can be used as a unique performance of the new selection film material and are widely used in molecular level filtration, and the PDLC in colour image display research has made some progress, in recent years, some new PDLC equipment has been introduced, such as the memory effect of the bile phase PDLC. Psychographic PDLC, etc. However, the contrast, dynamic response and viewing angle of these display devices still need to be improved. To improve the display characteristics of PDLC, people have begun to study the feasibility of polymer liquid crystals as a substrate material [12].

PDLC materials can also be used as smart glass. The smart glass here is not an architectural "glass system technology", but a glass that can automatically adjust the light transmission rate, thus effectively adjusting room temperature and reducing the load of air conditioning, to protect the environment and save energy. This technology has taken shape abroad and has been well used, but there are few domestic applications.

At present, PDLC materials have been used in re-intelligent glass, projection display devices. Asahi Glass Co. of Japan Ltd. And NHK have tested PDLC projectors based on TFT, respectively. Duane et al. made a large PDLC window. The French regiment Saint-Gobain has developed large-scale touch screens with an area of 3000 × 1000 mm and a thickness of only 11 mm, with temperatures ranging from -2000°C to 600°C, while their company has produced electronically controlled smart glass with a threshold voltage of 110V end excellent other photoelectric performance. Both types of products are now in production. Research Frontiers Plus in the United States has developed a large-area display of 24 ×1200 mm, using temperatures ranging from -40 °C to 120 °C [12-13].

6.1.12 The current research status of polymer dispersion liquid crystal

At present, the research stage of the laboratory has achieved some results.. From a domestic point of view, the field of electronic science and technology has seen advancements, particularly in the development of PDLC (Polymer Dispersed Liquid Crystal) membranes. By incorporating a crosslinking agent into the pre-polymer, researchers have achieved a threshold voltage of 10 V, with a saturation voltage around 20 V, which indicates the operational efficiency and responsiveness of these membranes. Ren Yan, Zhou Min, etc., improved by changing the intensity of ultraviolet radiation and film temperature, obtained a contrast of 103 PDLC films. The Spring Institute of Physics of the Chinese Academy of Sciences can control response times to up to 13 MS on T_{ON} and up to 35 by precisely controlling curing conditions, such as film temperature [13]. The Institute of Precision Machinery and Physics of the Chinese Academy of Sciences and the Northern Liquid Crystal Research and Development Center have solved the hysteresis effect by controlling the thickness of the membrane.

From abroad, Italy's Fiore Pasquale Nicoletta, Kinzie Caruso uses the PIPS-TIPS legal PDLC membrane with liquid crystal droplets of approximately 1-4 μm, with an open-state light through rate of 90%. The pass-through rate of off-state light is 19%. Ireland's L. Petti et al [14]. Use different types of polymers molecular enough to obtain threshold voltages of 2.4 V, TON up to3.5 standoff up to 0.22 Ms. Recent studies by Ms. Korea's S.J. and colleagues highlight significant improvements in PDLC films, achieving a contrast ratio exceeding 100:1. They report a threshold voltage of just 6 V and an impressive response time of less than 30 milliseconds. These advancements further enhance the operational effectiveness of PDLC technology in various applications.

6.1.13 The significance and innovation of the topic

As a new type of photoelectric display material, PDLC film material can be widely used in light valves, controlled intelligent glass flexible large screen displays and other fields, PDLC film material has attracted wide attention from scholars. Research in this area has also been carried out in-depth, involving the preparation of PDLC film materials, phase maps and electro-optical properties. This topic applies the UV curing legal system to prepare PDLC film materials. The effect of photo-polymerization monomer and preparation conditions on PDLC membrane was studied.

First of all, the sulfur-containing alcohol compound monomer was selected to make it soluble with the liquid crystal, and other polymeric monomers were added, so that sulfur-containing compounds and other double-bond polymerization monomers could be co-polymerized, a polymer network was formed through ultraviolet light exposure, the PDLC membrane containing thiol compounds was prepared, and the PDLC membrane containing thiol compounds was compared with the PDLC membrane without thiol compounds.

Conclusion

Through the comparison of PDLC membrane network structure and electrical-optical performance produced by the UV1/LC system and the UV2/LC system, the change of network pore size in the UV1/LC system is not obvious, and the change of UV2/LC system is more obvious, the analysis may be the reasonUV1/LC system contains sulfuryl polymerization monomer, this experiment selected sulfur hydrogen functional group of compounds, caused by the generation of more free radicals, and other polymeric monomers polymerization, and because of the large molecules containing thiol, molecular movement is not obvious, so that the polymer network shrinks not obvious when the temperature changes, so that the mesh changes little with temperature, polymerization monomer polymer network than does not contain sulfur alcohol. The polymer network of the complex is dense. In addition, with temperature changes, PDLC membranes containing thiol polymerizing monomers have less effect on the pass rate of light in different bands. Adding sulfur-containing alcohol compounds to the PDLC membrane can improve the thermal stability of the PDLC membrane.

References

Kurihara S, Nakatake Y. **(2000)** Optical shutter driven photochemically from super molecular polymer/liquid crystalline networks. *Polymers for Advanced Technologies*, 8- 12(11): 434-441.

Peng S C, Yu J W, Lee S N. **(1997) Effect** of droplet size on the dielectric properties of PDLC films. *Journal of Polymer Science Part B: Polymer Physics*, 7(35): 1373-1381.

Pogue R T, Natarajan L V, Siwecki S A, et al. **(2000)** Monomer functionality effects in the anisotropic phase separation of liquid crystal. *Polymer of advanced chemistry, 41*(1): 733-741.

Ren Hongwen, Huang Xixuan, Yu Xibin. **(1997)** the polymer disperses the liquid crystal phase separation and the photoelectric properties of its membrane. Functional materials, 28 (6):662-664.

Talroze R V, Zubarev E R, Kuptsov S A, Merekalov A S, Yuranova T I. **(1999)** Liquid crystal arylate-based networks: polymer backbone-LC order interaction. *Reactive Function Polymers,* (41):1-11.

Henry R M, Ramsey R A, Sharma S C. **(2004)** Effects of Crosslinking Agent, Cure Temperature, and UV Flux on the Electro-optical Properties of Polymer- *Dispersed Liquid Crystal Cells.J Polym Sci Part B: Polym Phys*, ,42: 404–410

Byung K K, Young S O. **(1994)** Morphologies, and electro-optic properties of polystyrene/liquid crystal composite films. *Journal of Polymer Science Part B: Polymer Physics*, 32(3):561-567.

Kinya K, Keiji T, Munekazu D **(1998). Reflective** liquid crystal color display technologies. Electronics and Communications in Japan *(Part II: Electronics*), , 81(8):32-40.

Almeida P L, Tavares S, Martins A F, et al. **(2002)** Cross-linked hydroxy propyl cellulose films: mechanical behavior and electro-optical properties of PDLC type cells. *Optical materials*, 20(2): 97~100.

Song J H, Park L S, Eeon E T, et al. **(2003)** Functional Materials and Electro-optic Prosperities for Polymer Stabilized Ferroelectric Liquid Crytal. *Synthetic Metals*, 137: 1411~1412.

Talroze R V, Zubarev E R,Kuptsov S A, Merekalov A S, Yuranova T I **(1999)** .Liquid crystal arylate-based networks: polymer backbone-LC order interaction. *Reactive &Function Polymers*, (41):1-11

Gill G, Pojman N, Willis J A **(2003)** Whitehead J Polymer-dispersed liquid crystal materials fabricated with frontal polymerization. Journal of Polymer Science Part A: *Polymer Chemistry*, 41:204-212.

Cristina E H, María J G, Patricia A O, Roberto J J W. **(2003)** Polymer-Dispersed Liquid Crystals Based on Polystyrene and EBBA: Analysis of Phase Diagrams and Morphologies Generated. Macromolecular *Chemistry and Physics*, 204(7):928-935.

Ren Hongwen, Huang Xixuan, Yu Xibin, etc. **(1997)** The optical properties of the polymer dispersion liquid crystal phase separation and its membrane Functional materials, 28(6):662-663.

Ulrich M Xavier. **(1995)** Electron beam processing for polymerization induced phase separation: preparation of polymer dispersed liquid crystal films. *Nuclear Instruments and Methods in Physics Research B* 105, :262-266.

Pogue R T, natarajan L V, Siwecki S A, Tondiglia, Sutherland, Bunning T J. **(2000)** Monomer functionality effects in the anisotropic phase separation of liquid crystal. Polymer, 41:733-741.

Petti L, Mormile P, Blau W J. **(2003)** fast electro-optical switching and high contrast ratio in epoxy-based polymer dispersed liquid crystals. *Optics and Lasers in Engineering*, 39(3):369-375.

Zhou J, Petti L, Mormile P, Roviello A. **(2004)** Comparison of the thermo- and electro-optical properties of doped and un-doped MOM based PDLCs. *Optics Communications*, 231(1-16):263-271.

Mucha M. **(2003)** Polymers as an important component of blends and composites with liquid crystal.Prog.*Polym.Sci.*28, :837-873.